THE BIRD OF TIME

THE BIRD OF TIME

*

The science and politics of
nature conservation—a personal account

N. W. MOORE

The right of the
University of Cambridge
to print and sell
all manner of books
was granted by
Henry VIII in 1534.
The University has printed
and published continuously
since 1584.

CAMBRIDGE UNIVERSITY PRESS

Cambridge

New York   New Rochelle   Melbourne   Sydney

CAMBRIDGE UNIVERSITY PRESS
Cambridge, New York, Melbourne, Madrid, Cape Town, Singapore, São Paulo, Delhi

Cambridge University Press
The Edinburgh Building, Cambridge CB2 8RU, UK

Published in the United States of America by Cambridge University Press, New York

www.cambridge.org
Information on this title: www.cambridge.org/9780521338714

First published 1987
Reprinted 1987, 1988, 1991
Re-issued in this digitally printed version 2009

A catalogue record for this publication is available from the British Library

Library of Congress Cataloguing in Publication data

Moore, N. W. (Norman Winfrid), 1923–
The bird of time.

Bibliography
Includes index.
1. Nature conservation – England. 2. Nature
conservation – Great Britain. 3. Moore, N. W. (Norman
Winfrid), 1923–   . I. Title.
QH75.M66 1987   33.95'16'0942   86-28379

ISBN 978-0-521-25259-1 hardback
ISBN 978-0-521-33871-4 paperback

CONTENTS

*

TO JANET

ACKNOWLEDGEMENTS

───────────────── ＊ ─────────────────

First, I should like to thank all my fellow workers in the conservation movement, whatever their organisation. Many of the ideas described in this book have arisen from discussions with them and have been tested in our joint enterprises. I should like especially to pay tribute to my colleagues in the old Nature Conservancy and in its successor, the Nature Conservancy Council. It has been a lifelong happiness to have worked with men and women for whom conservation was much more than a job, and whose enthusiasm has not been diminished by having to work through the system. Their friendships mean much to me.

Sometimes in this book I have been critical of certain actions and certain organisations. I should like to assure those whom I have opposed that much of the conflict has been thoroughly constructive. I am warmly indebted to those, who, despite differences of opinion, have remained friends. I hope that I shall make it clear that the future of conservation depends on integrating it effectively with those other interests which they have defended: the final goal transcends our separate endeavours.

In writing this book I have discussed details with many colleagues in the Nature Conservancy Council and the Institute of Terrestrial Ecology. I am most grateful for their help and I hope that they will forgive me for not mentioning them name by name: they are too numerous for me to do so. Needless to say they are in no way responsible for any of the errors which may occur.

I am indebted to the editors of the *Journal of Ecology* for permission to reproduce figures 13, 14 and 26, to the editors of the *New Naturalist* Series of William Collins and Sons for permission to reproduce figure 16, and to the editor of *British Birds* for permission to reproduce figures 32 and 33.

This book is for biologist and layman alike, and so I have sought criticisms from members of my family as representatives of both. I should like to give special thanks to my wife, Janet Moore, and my son-in-law, Richard Cohen for the immense amount of time and care they have given in criticising the manuscript. Their help has been invaluable. I would also like to thank Donald Watson for providing the frontispiece. Finally, I thank Zoë Conway Morris for typing this book so efficiently and for cheerfully putting up with the dreadful idiosyncrasies of my handwriting.

and a notice saying 'Keep Out', while 'Conserve' suggests positive management of the land so that its plants and animals are actively helped to survive.

Unfortunately the word conservation is used for the treatment of works of art and buildings as well as for looking after species and habitats. Human artefacts do require positive management in the form of repairs, but living things need much more: they require those conditions under which they can survive and perpetuate their kind by reproduction. There is a world of difference between the two activities, and it is a great pity that the same word has to be used to describe both. When there is any doubt it is best to describe the maintenance of living organisms by the phrase 'nature conservation', but this can be cumbersome, and in this book, whose subject is confined to nature conservation, I shall always use the single word conservation to describe it.

Like ecology, conservation has had its meaning extended. We talk about the conservation of a species or a habitat, meaning their safeguard and maintenance, but the word is also used as a collective term to cover the whole range of activities connected with conserving nature, in the same way that the word agriculture can be used to cover much more than the cultivation of fields.

One of the conclusions of this book is that conservation should be a universal aim; nevertheless one does need a collective noun to describe those whose special role, job or inclination is to conserve nature. Conservationist is an ungainly word and already has developed overtones which are damaging to serious down-to-earth conservation advocated here. I apologise for its use but, at present, there is no better alternative.

Habitat originally meant the particular environment inhabited by a particular species. For example, heather moorland is the habitat of the Red Grouse: a species of bird which feeds largely on heather and cannot survive in places where this plant is absent. Many plants and animals are not so exactly constrained, but they are restricted to living in woods, or lakes or meadows etc. More recently the word habitat has come to be used to describe such places. In other words the emphasis has been taken away from the requirements of particular species and put on the general types of environment which support them.

WORDS

--------------------------- * ---------------------------

One of the difficulties in writing about conservation is that the words which we have to use most often – conservation itself, ecology and habitat – are all inadequate. This is partly because they have changed their meanings in response to the events described in this book.

Words matter because they have far reaching effects on the way we think about things; therefore I must make it very clear how I use these key words.

Ecology was originally a technical term to describe that branch of biology which is concerned with the study of the relationships between plants and animals and their environment. Ecologists often became concerned about the conservation of what they studied, and as a result the word ecology has developed over-tones which it previously never had. Today it can refer to a way of life or political persuasion, based on ecological concepts; we have an Ecology Party. It is unfortunate that this useful word has been extended so far that its original meaning has become obscured. In this book I shall always use the word in its original sense to describe the particular scientific discipline which deals with living interrelationships.

Conservation has also acquired new overtones. That excellent naturalist, the late James Fisher, used to say 'Preserve or Conserve – I don't care which: it's all jam to me'. Most conservationists would now disagree with him. Originally the two words were virtually synonymous, but today preservation has developed a negative ambience and conservation a positive one. 'Preserve' suggests putting a fence round the land concerned

Strictly these places should be called biotopes, but since this word remains a technical term, I shall follow common usage in this book and call woods, heaths, etc. habitats. It should be remembered that they are easily recognisable types of country each of which supports a characteristic range of species, owing to the conditions which they provide.

The idea of habitat is unfamiliar to us because humans are no longer constrained by it. In contrast, all wild plants and animals are restricted in their numbers and distribution by their habitat requirements. Therefore the idea of habitat is crucial to the understanding of conservation.

FIGURES

*

This book is about ideas. The simple pen and ink sketches are included as a reminder that conservation is about actual plants and animals and actual places which are enjoyed, studied and used by actual people. Most of them are of the less familar species mentioned in the text. Wherever possible I have drawn them in the field, or from photographs taken in the field.

INTRODUCTION

———————————— * ————————————

Conservation is about the future – what we do today about our living resources will have profound effects on future generations. Recognition of this fact and of its urgency is a very recent growth. The profession of conservation biologist is a new one: it owed its birth to the formation of the Nature Conservancy in 1949. This book describes the experience of one man fortunate enough to be a pioneer in this new form of applied biology. Just because the circumstances and the profession were new, we were forced to look at conservation problems in their simplest and starkest forms. Therefore it may be easier for those of us who were in at the beginning to see the conservation movement in perspective and to appreciate how important it is.

Conservation matters; that has been much the most important conclusion which I have drawn from my own experience. I have seen conservation develop from a fringe activity to an essential element in human development. Today I would go so far as to say that the search for economic prosperity, democracy and peace will fail unless the importance of conservation is perceived and taken into account in all these endeavours. Whatever technological advances are made, the husbanding of living natural resources will always be necessary for our physical and spiritual needs. Therefore our future well being will depend to a very large extent on what we do to our environment.

These are large claims and are not yet accepted by most people. A glance at political manifestos and at the editorial and business columns of newspapers shows clearly that most do not perceive the extent of human dependence on nature. If it is mentioned at all, nature is usually relegated to 'Nature Notes' and whimsical articles about Natterjack Toads. Conservation is made to look frivolous, sentimental and backward looking, opposing development and more concerned with the past than the future.

How has this extraordinary and dangerous state of affairs arisen? I suspect it is partly because conservation is both a complicated concept and a new one. Conservation is both a subject and an aim. It involves fundamental and applied science, technology, economics, administration and politics and requires the understanding of people and society. It is motivated by strong feelings as well as by objective reasoning. The same can be said of two other great realms of applied biology – medicine and agriculture. The crucial difference is that the goals of medicine and agriculture can be expressed in very simple and understandable terms – healing the sick and feeding mankind. By contrast, the goals of conservation are complex: it is part or should be part of numerous human activities, not least of medicine and agriculture. This makes it difficult to recognise.

Conservation only became crucial with the explosion of the world's human population and with the industrialisation of agriculture. In the recent past, conservation was indeed less urgent and could be ignored; so it did not form part of ordinary thought and accepted wisdom. At least among older people, conservation is ignored simply because it is unfamiliar.

I believe that the main reason why conservation is not understood and why its importance is not perceived is because we fail to see it within the context of time. Only in terms of evolution can conservation be understood; only by looking at recent history can the magnitude of the new threats to the environment be appreciated. It is only by thinking in terms of evolutionary time that we can perceive the newness of our predicament and gauge the unique responsibilities of the generations alive today.

Since this book is based on the experience of the author, I should begin by trying to explain how I became a conservationist. Despite all the advantages of introspection I do not find this an easy task. I never met my paternal grandfather, who died just before I was born, nor my uncle who was killed in the First World War, yet both were naturalists with a particular interest in birds. My grandfather was a physician, but I suspect that he would have been a biologist if careers in biology had been possible in the mid-nineteenth century. My father also became a medical man but, not unnaturally, reacted against so much ornithology in the family. Instead he became an authority on the development of sailing ships and their rigging. As soon as he could, he took me to see the ships at Newhaven (the nearest port to Lewes, where we were living at the time). He was amused, but not wholly surprised, that the only things I looked at were the gulls.

Much of my school days was spent enduring classwork and escaping to the fields whenever the slightest opportunity arose. Escape was not always legal. One of my happiest early memories is of my first encounter with a Grey Phalarope. I sat entranced, as I watched it spin round on the surface of a little pond only a few feet away. The fact that I had no business to be in the Hampshire countryside at that moment, I am afraid, added to my joy. When I took final leave from the headmaster of my next school, he looked at me quizzically and said 'Moore, yes . . . the boy who spends half his time drawing and the other half at the local sewage farm'. There was a good deal of truth in that.

My formal education gave me virtually no biology until I went to university. This meant that I never did learn the nerves of the dogfish. I had the huge advantage of never being bored by my favourite subject. My love of nature was fostered by my parents. Neither was a naturalist, but, from my earliest days, they encouraged me to be one and to take natural history seriously. They introduced me to people who were very good naturalists, notably Hugh Whistler, the authority on Indian birds, and Maud Brindley (neé Haviland) who had travelled widely in Siberia before the Revolution, and was a very good all-round naturalist. Thus the basis for my life in conservation was an immense and, probably, innate delight in nature and the wish (part selfish and part unselfish) to see the source of that happiness maintained. I was given confidence by meeting highly competent and practical people who felt that the natural world mattered. It took me many years to learn that it mattered fundamentally.

My generation and the environmental movement grew up together. The activity which we now call conservation went through two stages and is now embarking on a third. The first was the pioneer stage, the stage of prophecy. A few individuals observed what was happening in the world and advocated enlightened agricultural, forestry and fishing practices and the establishment of national parks and nature reserves in order to stem the growing destruction wrought by industrial man. The pioneer stage was followed by a stage of limited action. In Britain, the early advocates of nature conservation were successful in getting the Nature Conservancy formed and National Nature Reserves and National Parks established in the years following the Second World War. This official activity was followed by a considerable growth in the number of voluntary conservation organisations. Conservation became part of the national life, but on a very small scale. Essentially it was a service for a minority run by specialists. In the 1960s, those concerned with conservation saw clearly that effective conservation could only be achieved in the future if it received a much wider support from the general public and government. Today we are still at the beginning of this important shift from limited specialised intervention for limited purposes to general acceptance that conservation must be an integral part of human activity at all times and in all places. In time it will be seen that conservation has provided one of the most potent political ideas since Marxism, and that it is an idea that will tend to unite rather than divide mankind, but that time has not yet come.

My generation grew up in the pioneer stage of conservation, worked throughout the stage of limited action, and hopes to live long enough to see real progress in the third stage – that of universal acceptance of conservation. The rate of change is now so great that one person can experience a great deal of history in a lifetime. For example, I can just remember talking to an old man who described to me the firework display at the coronation of Queen Victoria in 1837. His sister gave me a little statue of the Duke of Wellington: she had seen him as a girl in Kent and assured me that it was 'the Duke to the life'. More relevant to this book, my grandfather, after whom I was named, was a protégé of Charles Waterton, the Yorkshire squire and explorer. It was he who set up the first nature reserve in Britain in 1821. Therefore I feel personally linked with the conservation movement from its earliest beginnings.

As a teenager I helped stack corn into stooks and learnt to milk a cow by hand. Thus I have personal experience of traditional agriculture, while today I am involved with a system of farming in which sophisticated business techniques and electronics are used as a matter of course. The last forty years have been a fascinating time for those of us privileged to have lived at the formative stage of the environmental revolution, and in that period to have earned our living by practising conservation.

This book describes that experience and draws conclusions from it. It is not a history of the conservation movement, nor is it an autobiography, but it contains elements of both because it records my observations on the way ideas about conservation have developed.

The future of conservation lies with us all and so this book is written for Everyman, although I hope that biologists and conservationists will find its details of special interest to them. I have tried to give an impression of what it felt like to be a conservation biologist in the early days of that profession.

The theme of time runs through the book: it is crucial. The first part of the book sets the modern scene in the perspective and context of time. The second and third parts recall my experience with the two subjects which have occupied most of my working life – the safeguard of habitats for nature conservation and the control of pollution. I was mainly concerned with site safeguard when I was the Nature Conservancy's Regional Officer for Southwest England (1953–60) and when I was the Nature Conservancy Council's Chief Advisory Officer (1974–83). My main work on pesticides and other pollutants was done in the period 1960 to 1974 when I was in charge of the Nature Conservancy's Toxic Chemicals and Wildlife Section at Monks Wood in what was then Huntingdonshire.

Scotland, the north of England and Wales have contributed greatly to the conservation of Britain as a whole, and my routine duties frequently took me there. However, most of the action described in this book took place in the southern half of England. That is not because it was more important from the conservation point of view, but because it suffered the main threats from habitat destruction and pollution during the period under review and because it was where I was based.

In the fourth part of this book I draw the threads of experience together and discuss their implications for the future, outlining what I think should be done.

The achievements of conservation scientists and administrators in the last forty years, though limited, were fundamental and could not have been successful without the commitment of a quite small number of individual men and women. As this second stage of the environmental movement draws to a close the commitment of the few has to be transmitted so that it becomes the commitment of the many – that is today's challenge. How can we act in time?

PART I

* * *

Time and conservation

PART I

1

*

Evolutionary time

Since conservation is about the future, it should be one of the main concerns of mankind. Success or failure in conservation today will impinge increasingly on economics and politics and will affect human happiness. The need to maintain sustainable yields from our living resources should have introduced a new element into classical economic and political thinking, but it has not done so because conservation is still considered to be trivial and peripheral. Conservationists cannot expect to affect economic and political decisions unless they can show that conservation really matters, that it is truly concerned with our future. We have failed to put it in its true perspective. As already suggested, that perspective is one of time. Conservation makes little sense unless its aims and practices are related to development in the future, in other words to evolution.

The timescale of evolution is so different from that of ordinary life that it is not at all surprising that it does not impinge on most people's thoughts or actions. Most of our activities are geared to the day or the year. A few (such as the life of a parliament or the holding of offices, posts and jobs) extend for a few years, and some of our activities are related to our generation time and our lifespan; yet few of us live more than a thousand months, which is an extremely short period in evolutionary terms. By contrast, evolutionary processes are measured in millions of years. It is natural to feel that the evolutionary side of things is too big to be considered along with meal times, school terms or even marriage and life insurance. Indeed, for most of his history man has been totally unaware of his position in time relative to the earth's history, and even today most people who are aware of it ignore it. This is tragic, because for the first occasion in history, we can profoundly alter the course of future evolution and hence the happiness and well being of our own species in the future.

Biologists cannot ignore evolution so easily. Until recently all teaching of botany and zoology was based on a framework of evolutionary relationship. Today so much emphasis is put on molecular and cellular mechanisms that evolution can momentarily be forgotten but, in the long run, neither gene nor cell nor species makes sense except in terms of evolutionary development. It is the essence of biology. Much of what is taken for granted by biologists has to become common property if conservation is to be understood and achieved.

A biologist with a grounding in geology cannot help seeing the world differently from someone without that training, because the dimension of time pervades what he sees and adds interest and delight to it; he can see the past in the present.

I often walk along the disused railway line which runs through the fens of the village in which I live. Horsetails (Fig. 1) flourished

Fig. 1. Migrant Hawker (*Aeshna mixta*) (a) and Field Horsetail (*Equisetum arvense*) (b). Very similar dragonflies and horsetails were contemporary with the dinosaurs.

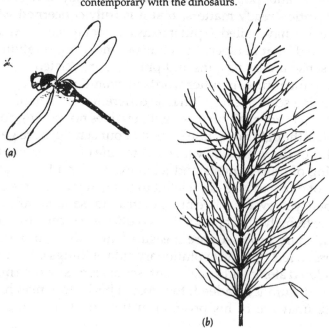

(a)

(b)

millions of years before the dinosaurs dominated the earth and there they are still, pushing their way up through the ballast of the railway track, showing in miniature what a forest of the Carboniferous period must have looked like. The railway line is flanked by large bushes that make it a sheltered place, which is very suitable for hawker dragonflies (Fig. 1). They fly to and fro, catching small flies and moths. The dragonflies were contemporaries with the dinosaurs. Insects like them must have perched on the warm backs of those huge creatures as they basked in the sun. The dragonflies are not as old as the horsetails, but they are immensely older than the flowering plants (like the buttercups growing in the meadows by the railway) and the flies and moths on which the dragonflies are feeding. Knowing these things, the biologist can see the past encapsulated in the present; this helps to put conservation into perspective. Even so, no one (biologist or not) can experience the huge stretches of past time in the same way that he or she experiences subjectively the time of day, season and lifespan.

Evolutionary time does become a little more imaginable when we can relate it to human history. The English Channel was only formed about 7000 years ago, that is some time after the invention of writing in Sumeria and Egypt. Yet the effects of separating British populations of animals from those on the continent can be easily discerned by the naturalist today. They are particularly marked in the Wagtails. The Pied Wagtail of Britain is conspicuously darker than its cousin, the White Wagtail on the continent (Fig. 2), and the British Yellow Wagtail has lost the blue head of its continental form. These changes have occurred during the span of about 200 to 300 human generations. We can begin to picture that.

Fig. 2. White Wagtail (*Motacilla alba alba*) (a) and Pied Wagtail (*Motacilla alba yarrellii*) (b). Recent evolution: there is now a clear difference between the populations either side of the English Channel, which was formed about 5000 BC.

(a)

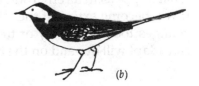
(b)

Some physiological adaptive changes have been even more rapid. For example, numerous insect pests have developed resistance to DDT during the 40 years or so that this insecticide has been used, and some populations of rats have become resistant to warfarin in the same period. Thus, if we look, we can see evolution working in history, and even within the lifespan of an individual. We can observe evolution occurring; it is not just something we read about in text books.

Much has been written about evolution, but the special relationship between conservation and evolution has been little discussed until recently. In this chapter I will outline those points which are basic to understanding conservation activities in the United Kingdom.

The world has existed for about 5000 million years and life for at least 3000 million years. The first vertebrates appeared about 550 million years ago, the first mammals about 200 million years ago and the first of our species about 1 million years ago. At any one time the world's plants and animals are made up of populations whose members can interbreed – we call these interbreeding populations 'species'. There are at least 2 million at present, perhaps 10 million or even more. However, if we could watch evolution occurring as a process, it would appear more like a river than a succession of separate paving stones on a road; we would see how one 'species' turned into another. Nevertheless, parts of the river would run into the sand and disappear. Extinction of species and whole groups is as much a part of evolution as the formation of new species. For millions of years the seas of the world were full of trilobites and ammonites (Fig. 3). They became extinct and left no descendants, but relations of the trilobites evolved into the crustacea which we know today, and relatives of the ammonites into the squids, octopuses and the nautilus of our present seas. The dinosaurs became extinct, but some of them evolved into the birds which are now one of the dominant components of today's living world.

Whether a species survives or becomes extinct depends upon what happens in its environment and on its own genetic fitness. It will become extinct if the environment to which it is adapted changes too extensively or too quickly. The degree to which it can adapt will depend on the amount of variation its population contains and, since large populations usually contain more variation than small ones, the chance of survival of a species is increased if the space it occupies is large. In other words common species are more likely to survive than rare ones though any species may be exterminated suddenly by a disease. Whether a species survives by adapting successfully to change in the environment, or becomes extinct, always depends on the interplay between the species itself and its environment.

Geological studies show that both gradual and rapid environmental changes have occurred in the past. Large-scale environmental changes are likely to cause an increase in the rate of extinction. There is fossil evidence that the great upheavals which occurred at the end of the Permian and Cretaceous periods, and more recently in the Quaternary Ice Age, caused massive extinctions even though they probably took thousands of years to develop. The much more rapid changes which man is now making on the world environment by destroying rain forests, making deserts and polluting the shallow seas are likely to produce effects as momentous as any produced by nature in the past.

It is harder to realise that the changes in agricultural practice which we are experiencing today are no less catastrophic in their

Fig. 3. A trilobite (a) and an ammonite (b). Both belonged to extinct groups which were once very numerous.

(a)   (b)

effects. Yet, like the destruction of rain forests, they are occurring at a rate which prevents the vast majority of wildlife species on farmland adapting to them. To take a homely British example, the caterpillars of Blue butterflies live on various species of vetch, which have been components of natural grasslands and the meadows of traditional agriculture for thousands of years. Since the Second World War most of these meadows have been replaced by grass leys consisting of monocultures of Italian Rye-grass. The Blue butterflies have not been able to switch from feeding on vetches to feeding on Italian Rye-grass. Considering the complicated change in physiology and behaviour that such a change would require this is not surprising. Even the Brown butterflies and the Skipper butterflies which feed on other species of grasses have failed to make the switch. Thousands of other species have been affected by changes in agriculture in this way.

There is a fundamental difference between previous catastrophes and that which man is causing today; man has it in his power to modify the one he is causing and to reduce its harmful effects. If man fails to use his intelligence in this way, the world will experience an unprecedented extinction rate due to the scale and speed of his activities. As a result, the future of evolution on the earth will be immeasurably restricted and man cannot fail to be the loser if this occurs. Thus we can conclude that the most fundamental aim of conservation in the years to come is to maintain a world in which evolution can continue. This is not a trivial objective!

When we look at conservation on an evolutionary time scale we are forced both to recognise its importance and to see the relevance of technological change. We have to consider the long-term strategy of conservation.

We have to recognise that all conservation aims are relative, not absolute, since the world will become uninhabitable in 5–6 thousand million years time. Sir Otto Frankel, one of the authors of the important book *Conservation and Evolution*, believes that all conservation objectives should contain a time-scale. This is a useful approach because it helps us to determine our priorities. There are many conservation objectives which are less fundamental than ensuring the continuance of evolution on the grand scale, yet they are worth pursuing. For example, some species are now so rare and their requirements so demanding that we cannot hope to do more than enable a few more human generations to study and enjoy them. Other species may be kept going over very long periods of time by the judicious use of botanical gardens and zoos. This might be well worth doing even if the species have no role in providing sources of new species in the wild in the future. These are laudable limited aims, comparable to the conservation of buildings and works of art which can be preserved for centuries but not indefinitely. By assessing conservation projects in relation to time in this way, we are continually forced to consider the fundamental purposes of conservation.

I have often sat in railway stations and watched people go by, observing them as individual children and grown-ups with different jobs, different backgrounds. It is easy to forget their relationship with time. Yet if one does stop, to think of them in this way, the scene acquires much more significance. One sees each person as a traveller, not only on a particular railway journey on a particular date, but in life itself. The ordinary becomes marvellous and one is led to consider the destiny of the people one is watching.

A similar situation occurs when one sits in a field in the countryside: one can observe plants and animals simply as individual objects, but when one relates them to time, a wonderful evolutionary journey is revealed. One cannot escape the conclusion that their future matters.

Human evolution will be the product of our genes and the natural and the cultural environments which we make for ourselves. Whether we consciously guide our own evolution or not, we shall always be affected by the other forms of life with which we share the planet, but they will depend increasingly on our decision making. When conservation is put in its evolutionary context we find ourselves challenged by new and awesome responsibilities.

2
*

Human time

The portion of time experienced by mankind is only a minute part of evolutionary time, but it is immensely important in evolutionary terms because it covers the development of the one species which will control the future of evolution. It is only by looking at the effects which man has had on the flora and fauna in the past – and particularly in the very recent past – that we can see why the situation today is unprecedented and why we have reached a critical point. We should see why man's activities are already as biologically significant as those forces which caused the gigantic geological upheavals in the past, and could become even more destructive than them.

What has happened in Britain can be taken as typical of what has happened in many other parts of the world. In this country there is still a wide range of plants and animals. There are species which can live only in woods, others only in marshes and others only in grasslands etc. Together they make up a complicated hotch-potch and we can only make sense of it by studying the past.

First we have to discover what the habitats were like during the long period when man was too rare a species to have any significant ecological effects, because the original habitats will explain the origin of most of the species which we have today. Then we have to study the effects which man had on those habitats when he became more numerous, because that will explain the present population sizes of our different species.

For most of his existence, man has been a hunter gatherer; for most of the short time that he has been a farmer, he has kept no written records. Therefore what we know about the effects of his activities is largely based on the study of subfossil pollen and wood related to human artefacts. When the ice withdrew, about 8000 BC, man must have been a rare species in Britain. He must have had local effects on habitats by using fire, and he may well have affected the numbers of Reindeer and the Great Elk by hunting, but he did not possess the tools with which to modify the natural changes from tundra to birch and pine forest and from pine forest to deciduous forest. The only open lands were marshes, dunes, sea cliffs, the highest mountain tops, and grasslands maintained by wild cattle and deer in places where the forest cover had been broken by wind blow and fire. All else was forest.

The formation of the English Channel in about 5000 BC prevented many continental species from colonising Britain. Britain, like all islands, has a less diverse flora and fauna than that of the neighbouring lands on the continent.

While the natural habitats of Britain were still adapting to the drier sub-boreal climate which followed on from the Atlantic period, man gradually became a farmer. He first domesticated the dog to help him hunt. Then hunting areas (perhaps consciously conserved as such) became pastures for increasingly domesticated cattle, sheep and goats, and areas treasured for wild food plants became croplands where seeds of some of the plants were deliberately sown to provide for next year's food. We shall never know the exact details of what happened, but it is clear that the domestication of plants and animals developed together to mutual advantage and, by about 1000 BC at the latest, most of the human inhabitants of Britain could be called farmers. The forest had to be cleared to provide land for crops and grazing (see Fig. 4). This was difficult until metal tools had been invented, and so the first farms were on thinner soils, where trees were smaller and easier to cut down. Therefore the land on the chalk and poorer sandy soils, and where tree growth was checked by flooding, was cleared first. By the time the Romans invaded Britain and recorded what they saw, a very large part of lowland England had been turned from forest to farmland. Then, as a deliberate colonial policy, the Romans opened up the country, settling retired soldiers in the remoter border areas. A network of roads was organised, forests cleared and huge areas of the marsh drained. Iron works were established in the depths of the Weald. By the end of the Roman period much of lowland England must have felt open and the remaining forests were already 'islands' in a sea of agricultural land. However, some of the 'islands' were very large and supported bears, wolves and boars and, possibly, the last wild cattle. Much woodland was already being managed for the production of wood and timber.

The Anglo-Saxons carried on where the Romans left off. The relative extent to which they cleared aboriginal forest, or secondary forest which had developed when the Romans withdrew, or simply dispossessed British farmers is not known. By the time the Normans took over the government of the country, the Saxons had founded practically all of today's towns and villages. Subsequent generations have done little more than dot the 'i's and cross the 't's of forest clearance in England. The great forests of Wales, Scotland and Ireland were mostly cleared in mediaeval times. However, as late as 1726 the 'Several Hands' who wrote *A Natural History of Ireland in Three Parts* could say that 'There are still great woods remaining'. Even in Leinster itself, some of the woods were 'many miles long and broad'. They were not to last

Fig. 4. Great Wood, Creech Hill, Dorset. This chalk hill retains its forest cover on the north-facing scarp. Its south-facing scarp was cleared in ancient times and is now chalk grassland. The flat top is cultivated.

long. By the beginning of the present century, only about 4% of Britain (and less of Ireland) was still covered by woodland.

The loss of timber for shipbuilding was decried in the sixteenth, seventeenth and eighteenth centuries and plantations on open ground were made, notably in Scotland, in the eighteenth century. Large-scale planting of trees – mainly conifers – did not begin until the state intervened by establishing the Forestry Commission in 1919. Today more than half our woodland consists of conifers planted in the lifetime of the older inhabitants of the country.

For species of wildlife dependent on woodland habitats, the loss of woodland was to some extent made good by the creation of hedges. At the time of their greatest extent, hedges covered about 1% of the country (an area equivalent to about a quarter of the remaining deciduous woodland). In Saxon England, many hedges were made by leaving strips of woodland uncleared, but from mediaeval times onward they were mostly planted. They formed a huge network totalling about 998 000 km (620 000 miles), extending over all the lowlands – except for areas such as the Fens and Romney Marsh, where ditches made better barriers. In many cases hedges linked the remaining woods and plantations together. Many hedges have been destroyed during the last 40 years, especially in areas where arable farming has supplanted mixed farming (see Chapter 4); however, nearly 805 000 km (500 000 miles) probably remain.

Hedges of sorts are found in most countries in the world, but they have been planted more in Britain than in any other country. They have been part of the English scene for centuries and, since the enclosures from the sixteenth to the nineteenth centuries, have been quintessential in our landscape. They have provided a great deal of habitat for our commoner plants and animals, and woodland corridors for species which cannot disperse across open ground.

The destruction of woodland has been greater in the British Isles than in most European countries; 20% of France is still covered by forest. Therefore the effects of forest destruction in Britain are likely to be proportionally greater. There are relatively few records of plants and animals before the seventeenth century and practically no systematic assessments of population numbers until the immediate post-war period. However, we have information about some extinctions and we can make some inferences about population sizes.

The reduction of forest cover from about 90% of the land surface to less than 10% must have had a roughly proportional effect on many species. For example, it is estimated that today there are about 35 000 pairs of Greater Spotted Woodpecker in Great Britain. This species is dependent upon large trees for food and nesting sites. Its original population was probably nearer 300 000 pairs. On the other hand, the Skylark, which is one of the very few bird species which has been able to benefit from agriculture, must have been a relatively rare species in pre-agricultural Britain (there were probably fewer than 100 000 pairs). Today there are about four million pairs in the British Isles. It is much harder to assess the changes in status of species like the Yellowhammer and the Whitethroat which are adapted to living on the woodland edge. There is less wood but probably more edge.

The destruction of forest has had two major effects which have interacted with each other. It has reduced the total amount of woodland habitat and it has broken up the originally continuous area into small fragments. Thus, populations of plants and animals which have already been reduced by habitat destruction have also become fragmented. This does not matter if the total population is fairly large and its members are good dispersers, because the separate populations will be genetically connected and variability (and hence viability) will be retained. If each small population becomes isolated, however, it will become inbred and will lose variability, and so be less able to adapt to changing conditions. Small populations are also more vulnerable to accidental extinction. By the mid-nineteenth century it had become fashionable to make natural history collections. When it was known that mammals, birds or conspicuous plants and insects had become rare, specimens became collectors' pieces and the extinction of rare species was hastened. Happily, in more recent times rapid declines have stimulated special conservation measures to halt or reverse them.

Ecological theory would predict that the first species to become extinct as the result of habitat loss and fragmentation would be large animals with relatively small populations. The known facts about forest species support that theory. Wild cattle had disappeared by the end of the Roman period at the latest, Brown Bears and beavers by the beginning of the Middle Ages, wild boars in the seventeenth century and the wolf in the eighteenth century. The only woodland bird known to have become extinct was the Goshawk – a large bird of prey – which became so in the nineteenth century. Recently, falconers' Goshawks have escaped and taken to the woods and a small wild population now exists again in Britain. It is interesting that the original population became extinct when woodland in Britain covered about 4% of the country and it is now able to survive when – thanks to new plantations made since 1919 – there is about twice that amount. It is possible that some other woodland birds became extinct but were unrecorded. However, if we look at woodland species which survive in the large forests the other side of the English Channel it seems unlikely that we have lost many bird species. It is possible that the Middle Spotted Woodpecker, the Black Woodpecker and the Hazel Hen (Fig. 5) and conceivably one or two other species were once native to Britain and became extinct before they were recorded.

Fig. 5. Middle Spotted Woodpecker (*Dendrocopus medius*) (a), Black Woodpecker (*Dryocopus martius*) (b) and Hazel Hen (*Tetrastes bonasia*) (c). Birds of the European forests which may have bred unrecorded in Britain before its forests were reduced and fragmented.

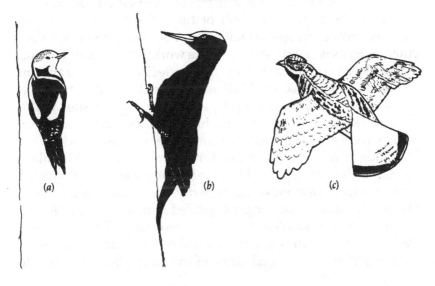

(a) (b) (c)

The destruction of marshlands was on the same scale as the destruction of forest. It led to the extinction of the Dalmatian Pelican, Crane and Spoonbill 300 years or more ago. The Bittern, Avocet, Ruff, Black-tailed Godwit, Black Tern and Savi's Warbler all became extinct in Britain in the nineteenth century. In the case of the last few species, extinction was hastened by collectors when populations had become much reduced. However, these birds have recolonised Britain from the continent in the present century and thanks to effective conservation measures have become re-established.

The Great Bustard was the biggest bird to exploit the rolling grasslands, produced by generations of English farmers. It became rare and finally extinct in the nineteenth century following a period when many of the grasslands were ploughed up to grow corn during the Napoleonic Wars.

Birds live at relatively low densities but they are good dispersers. It has recently been shown that the small isolated population of Red Kites in Wales is in contact with the much larger one on the continent and so is not threatened by inbreeding as was once feared. All in all it is not surprising that the great reduction of forest and marshland habitats has not caused the loss of many bird species.

Invertebrate populations are very much larger than those of vertebrates, and many species are good dispersers, therefore it is likely that we have not lost many of them either, despite the diminution and fragmentation of their forest and marshland habitats. However, species with poor dispersal powers are very vulnerable, as is clearly shown by the work of Dr Jeremy Thomas on the Black Hairstreak (Fig. 6). This butterfly is found in ancient woodlands in the south Midlands of England. Its caterpillar feeds on blackthorn, but it prefers old blackthorn bushes which are covered by the lichens in which it hides. Old blackthorn bushes were a characteristic feature of the oak-ash woods of the Midlands, which were coppiced on a long rotation. The Black Hairstreak has disappeared from many of the woods in which it once occurred. Doubtless this is partly due to the fact that most Midland woods are no longer coppiced and, as a result, the old blackthorns get shaded out and eventually die. However, blackthorn is an extremely common shrub and many thickets of it occur within the original range of the Black Hairstreak but do not harbour the butterfly. In his studies on it, Jeremy Thomas found that it was a poor disperser; apparently it has an instinct which prevents it from flying across open country. Doubtless this was valuable when the Midlands were covered by continuous deciduous forest but today it prevents gene flow between populations isolated by man and, since no wood in the Midlands is very large, most if not all populations are doomed unless conservation measures can be taken. Fortunately it is relatively easy to restock woods which contain blackthorn with Black Hairstreaks. This has been done successfully at Monks Wood

Fig. 6. Black Hairstreak (*Strymonidia pruni*) (a) and Silver-washed Fritillary (*Argynnis paphia*) (b). Butterflies whose population declines in Britain have been associated with the decline of the coppice and standards system of woodland management. Under modern conditions both have difficulties in recolonising woods that have become suitable. The Black Hairstreak is particularly poor at dispersing.

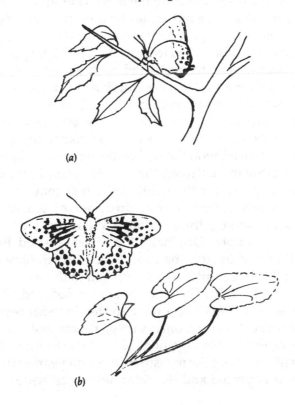

(a)

(b)

National Nature Reserve where the species was first discovered in 1828 and where it died out following clear felling in 1914–18. Black Hairstreaks have even been introduced successfully into woodland in Surrey outside its original known range.

An analogous situation to that described for the Black Hairstreak can also occur naturally – even with birds. For example Dr J. M. Diamond has noted how many of the forest birds of mainland New Guinea have failed to colonise the forests on small offshore islands, which appear entirely suitable for them. However, there is no doubt that man has greatly exacerbated the problem of genetic isolation by breaking up once continuous forest in many parts of the world and turning them into archipelagos of habitat islands.

In Britain, forest destruction has been particularly great and the remaining islands are particularly small. The disappearance of the woodland fritillary butterflies over most of the southeast quarter of England in recent years is probably due to adverse factors similar to those affecting the Black Hairstreak. The caterpillars of these attractive butterflies feed on wild violets, which are characteristic of the ground flora of recently coppiced woodland. When coppicing ceased to be profitable, violet populations in woods became greatly restricted and, as a result, fritillary populations became too small to survive. In those few woods where coppicing regimes were reinstated, violets returned in abundance, but not the fritillaries. They are now so rare in southeast England that the chance of one dispersing individual finding the coppiced woodlands has become very slight. Also, most of the farmland between the woods in southeast England is extensively sprayed with insecticides which may increase the hazards to the few fritillaries which do disperse from the remaining woods containing them. Monks Wood provides a well-documented example. Originally the Silver-washed Fritillary, the High Brown Fritillary, the Pearl Bordered Fritillary and the Small Pearl Bordered Fritillary all occurred in the wood. All had become extinct by 1970. The wood had been declared a National Nature Reserve in 1953 but, owing to the financial restrictions imposed on the Nature Conservancy, it was not possible to undertake extensive management of the reserve until the early 1960s. Thanks to energetic measures in recent years much of the wood is now coppiced and the rides have been widened, with the result that wild violets flourish. Yet, to date (1986) the reserve has been recolonised by only one species of fritillary (the Silver-washed, in 1984, see Fig. 6). Since there are very few fritillary colonies within 100 km, and spraying against cereal aphids and other pests is carried out extensively on surrounding land, this is not surprising. In this case it would seem entirely justifiable to reintroduce one or more of the lost species to ascertain whether current management practices can sustain viable populations within what is one of the largest remaining woodland 'islands' in eastern England.

Marshes and wetlands have always been ecological 'islands', and so the species which live in them have had to evolve effective dispersal mechanisms. Anyone who has dug a pond in his garden must be struck by the speed with which it is colonised by water beetles, water boatmen and dragonflies. Fresh water species in general have suffered a huge reduction of habitat by drainage schemes, but probably they have been much less affected by fragmentation of habitat than woodland species.

Grassland in western Europe must have also been 'island' habitats originally. Formerly they were found only on coasts, and in small areas by rivers where physical forces, sometimes aided by grazing animals, kept the trees at bay. Perhaps this is the reason why very small grassland habitats retain their flora and invertebrate fauna surprisingly well. For example, all of the once-extensive chalk grassland of South Cambridgeshire has been ploughed up except for two ancient Saxon defensive earthworks, the Devil's Dyke and the Fleam Dyke, the verges of a Roman Road, some railway cuttings and embankments and one or two chalk quarries, yet these tiny remnants still provide homes for nearly all the chalk plants recorded by John Ray in the seventeenth century. Pasque Flowers, Bloody Cranesbill and Juniper (Fig. 7) still occur in Cambridgeshire, and there are still Chalkhill Blue butterflies because there is still enough Horseshoe Vetch for their caterpillars. The remaining chalk grassland habitats of Cambridgeshire survived in the first place because they were too steep or difficult to cultivate. Rabbits kept them grazed, but when myxomatosis killed off the rabbits the grasslands began to be colonised by scrub and were seriously threatened. As soon as this was realised, the Cambridgeshire and Isle of Ely Naturalist Trust organised work parties to control the scrub, and, as a result, the last remnants of the Cambridge-shire chalk grasslands have been saved.

In historical times grasslands have fluctuated in area according to the relative profitability of corn and stock. Throughout the Middle Ages vast flocks of sheep were kept on the chalk downs and wolds as well as on the moorlands. Corn was grown on the clays; but when the Black Death struck in 1348–9 many areas could not be made to grow corn because there were not enough people to plough, plant or reap and so the area of grassland increased. Yet, by Tudor times, corn was being exported to the continent.

During those periods when Britain was at war in Europe it was 'up corn and down horn'. The Napoleonic Wars era and our own have seen vast inroads on the grasslands in order to grow corn. The Second World War, unlike the Napoleonic War, saw great changes in the nature of grassland as well as a diminution in its extent. Previously, when corn and turnips ousted sheep in East Anglia and the Cotswolds, the grassland which remained unploughed kept its ancient character. By contrast, many of the grasslands which had survived the Second World War and the

Fig. 7. Juniper (*Juniperus communis*) on the Fleam Dyke, Cambridge-shire. This Saxon earthwork is one of the few places where the chalk grassland flora survives in Cambridgeshire. It is now the only locality for Juniper in East Anglia. Scrub encroachment is being kept at bay by the Cambridgeshire Wildlife Trust.

years which followed it were turned into grass leys or were treated with selective herbicides. These killed the broad-leafed plants which made up a large proportion of many ancient grasslands. What superficially looked the same was really quite different. Not only were complicated societies of native plants replaced by monocultures of domesticated strains of grass, but the animal societies were changed too. For example, one of the typical plant species in unimproved pasture was the little vetch affectionately known as Eggs and Bacon; on it fed the caterpillars of the Common Blue butterfly (Fig. 8), one of the blue butterflies mentioned in the last chapter. Both plant and insect were exceedingly common. The vetch is eradicated by ploughing and

Fig. 8. Common Blue (*Polyommatus icarus*) (a), its food plant Bird's Foot Trefoil or Eggs and Bacon (*Lotus corniculatus*) (b) and the highly productive Italian Rye-grass (*Lolium multiflorum*) (c), which is widely sown in grass leys and has ousted the trefoil and hence the butterfly from thousands of fields in Britain.

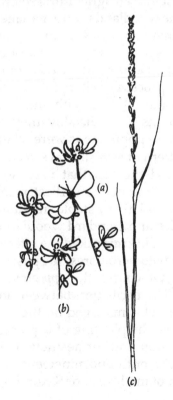

herbicides and so has become quite a local plant in many parts of England. With the vetch went the butterfly; the Common Blue is no longer common in much of England. In typical farmland country it survives in small colonies on wasteland where the vetch is still to be found, but some of these colonies are so small and isolated that they are always at risk. However, the species is not endangered in Britain because it is abundant on undercliffs and sand dunes, especially in the north and west. Sadly both plant and butterfly are no longer part of the ordinary experience of people living or working on ordinary farmland. The same story can be told of many other species.

We shall never know the details of ecological changes during the historical period, but if we relate archaeological and historical records to what we know about the requirements of species alive today we can piece together the general pattern with a good deal of confidence that it is correct. We know that from Neolithic times onward, man changed a country that was originally covered with forest into an agricultural sea with islands of small woods. Most of the woodlands were managed to produce wood and timber from native species. Similarly the grasslands were made up of native species as were most of the weeds of arable land. Only the crops and some of the domestic animals were alien species. The process of changing forest and marsh into a landscape of agricultural land with small woods caused relatively few extinctions – predictably they were of the larger animal species whose populations were relatively small.

The existing pattern of agricultural sea and woodland islands is very ancient, and the fact that very few extinctions were recorded from the time of the earliest records until our own demonstrates that a relatively stable situation had been reached – so long as coppice and standard woodland management and hay mowing and grazing mimicked the effects of lightning strike and the grazing of wild herbivores, the past continued smoothly into the present. Of course the *appearance* of the landscape changed greatly. The oscillation between grass and corn, the change from great fields managed on the strip system to fields enclosed by hedges, the planting of coppices and small woods for game conservation and for aesthetic reasons all made the country of the eighteenth and nineteenth centuries look very different from that of mediaeval or Saxon England. Even today, despite the enormous changes produced by modern technology much of the landscape looks quite similar to that of the eighteenth and nineteenth centuries if we half close our eyes and do not look at the small print. But it is the small print that is changing so drastically and rapidly, and nature conservation (unlike landscape conservation) is about the small print.

The huge changes we are now experiencing are of a kind that has not been experienced before. The coppice and standard system of management, which is probably at least 2000 years old, has virtually disappeared. Today most deciduous woods are either derelict or have been replaced by exotic conifers. Most of our grasslands are monocultures of Italian Rye-grass (Fig. 8), and traditionally managed grasslands, heathlands and wetlands have been greatly reduced and increasingly fragmented.

It is almost impossible to exaggerate the importance and significance of this recent, second agricultural revolution. It is occurring throughout the world. Unlike the other great upheavals which occurred at the ends of the Permian and Cretaceous periods and during the recent Ice Age, it is occurring within the time span of one human generation. Those of us who can remember the rural scene before 1939 have experienced the end product of something which had gradually evolved since the Neolithic period, and we have seen it changed by machines, oil, chemicals, plant and animal breeding and politics into an entirely new system. This period of dramatic change is both the cause and background of the case histories discussed later in this book.

PART II

---------- ★ ----------

The past: experience from conserving habitats

3
*

Disappearing heathlands

On my first visit, during the mid-1950s, the heath lay shimmering under a warm sun. A Dartford Warbler (Fig. 9) flitted hesitantly over the heather before rapidly disappearing into a gorse bush. The yellow flowers of the gorse filled the air with the warm smell of desiccated coconut. A lizard lay motionless in the sun and, when disturbed, it made a small rustling sound as it vanished into the shade. Hundreds of ants crossed the open sandy places, endlessly busy. The heath was ticking over in all its complexity, as it had for centuries.

On my second visit the whole area was black and some of the peat was still smouldering. A migrant Tortoiseshell Butterfly flew quickly across the burnt heath but did not stop.

Fig. 9. Dartford Warbler (*Sylvia undata*). A characteristic species of the Dorset heathlands.

On my third visit the heath was being ploughed; on my fourth, some months later, the heath had become a grass field and was being grazed by Friesian cattle. It was surrounded by a post and wire fence and the only sign that it had once been a heath was that a few gorse seedlings had germinated at the field's edge.

This type of experience has been a common one in our generations. It has been inevitable that many heathlands have been reclaimed, that many meadows have been ploughed up and resown with rye-grass, that many marshes have been drained, and many woods felled; yet, however great the economic gains may have been, the loss of such places has been significant. For a long time each has seemed an individual matter, to be mourned privately by individual people. Today we look at such events rather differently. The loss of each separate place is seen as an impoverishment of a valued total resource, something of greater importance that should concern us all.

How has this change of emphasis come about? We all had to learn by experience. It was the disappearance of heathlands which opened my eyes to the problem of habitat loss in general and to its economic causes, and it was the first of such cases to be documented. Therefore I shall describe it fully; but before doing so I must return to those agricultural developments which formed the background to my studies and I must outline the early history of the Nature Conservancy for whom I was working.

In the 1920s and 1930s the seeds of the second agricultural revolution were sown, but the effects were not obvious; most farmers did not plant grass leys or use herbicides or insecticides. Horses outnumbered tractors and farmyard manure was used more extensively than artificial fertilisers. In fact, agriculture was stagnant. It continued to support wildlife as it had for centuries. The threats to wildlife seemed to come from elsewhere. The absence of planning legislation and the low cost of land allowed extensive urban development on farmland throughout the country, notably in the particularly fertile areas surrounding London and along the English Channel coast. Most of Middlesex and large areas of Essex, Kent, Surrey, Sussex and Hampshire became built over. Urban development was the obvious threat to the countryside. In the 1920s and 1930s many people thought that wildlife suffered severely from the direct effects of human beings. Increased mobility did mean that more people from towns picked or uprooted flowers in the surrounding countryside. Despite the Bird Protection Act birds and birds' eggs were still collected extensively and rarer species were almost certainly affected. Population ecology was in its infancy and therefore few people thought deeply about what actually controlled the number of plants and animals; collecting was not put into context. Collecting was obviously bad for individuals and hence it was inferred that it was bad for the species and so it was given undue attention. Also, it appeared that one could do something about it: laws could be passed forbidding the taking of rare birds. As a child I used to read with fascination the notices fixed to trees and noticeboards which outlined the penalties for killing Glossy Ibises and other rarities. The countryside was full of these notices.

Every now and then some special habitat was threatened and efforts made to protect it. Sometimes the place concerned was turned into a nature reserve by the National Trust or the Royal Society for the Protection of Birds, but each was a special case, and the general threat was not perceived.

When increased agricultural productivity became a wartime priority and large areas of Britain were reclaimed most people still thought about conservation in the old way. The urban developer and the collector were the enemies. Most conservationists were not yet ecologists and were ill-equipped to perceive the new threats to wildlife.

The Second World War stimulated the revival of British agriculture; more surprisingly it stimulated the first stage towards a national conservation policy. There was something heroic about the formation of the Nature Reserve Investigation Committee in the darkest days of the war. Its work did not stem from recognising that a revitalised agriculture would pose a threat for conservation, but from a determination that conservation would be part of the better post-war Britain of which we dreamed. I first heard of the Nature Reserve Investigation Committee from my mentor and neighbour Hugh Whistler, who had retired from the Indian Police and was living at Battle. We discussed what was in effect, an embryo list of SSSI (Sites of Special Scientific Interest, see p. 59) for East Sussex. I remember that we agreed on the inclusion of the Rother Levels between Bodiam and Northiam. In those days the land there was poorly drained and marshy pools held water throughout the summer and provided habitat for Shoveler, Garganey, Redshank and Snipe. Also I had recently discovered the first Sussex locality of the Scarce Emerald Damselfly (see Fig. 17, p. 54) in an overgrown ditch on these levels.

The main conclusion of the Nature Reserve Investigation Committee was that 'The Government should take formal responsibility for the conservation of native wildlife, both plant and animal'. It showed how this could be done by establishing National Parks, National Forest Parks, Conservation Areas, and National and Local Nature Reserves. It also recommended the setting up of Geological Monuments.

Wartime training took servicemen and women to some of the finest country in the United Kingdom. We were grateful that our fares had been paid, for there were gaps in our military training when those of us who were naturalists, could go and watch birds, collect insects and look for plants. My own training as a gunner took me to Suffolk, Lincolnshire, the flats of Foulness, the Yorkshire Dales, Loch Fyne and the Cairngorms. All were places I had never visited before. We learnt at first hand what a varied and marvellous country Britain was. For those who were already conservation-minded it was wonderfully heartening to know that steps were already being taken to ensure that nature conservation would be part of the scheme of things in post-war Britain.

Immediately after the war, the Wildlife Conservation Special Committee was set up by Parliament. It acknowledged its debt to the Nature Reserve Investigation Committee in the remarkable White Paper entitled 'Conservation of Nature in England and Wales' (Command 7122). This far-sighted document was written between 1945 and 1947 and was published in 1947. It, and its Scottish equivalent (Command 7235), provided the blueprint for nature conservation in Great Britain.

The authors of the report were well aware that places of great scientific interest were being lost and that 'hurried and some-times ill-considered plans for agricultural expansion' was one of the causes. Reading the White Paper today, it is clear that its authors had no inkling that what had been done agriculturally as a war measure would be continued and expanded after it. Yet, in the very same year as the White Paper was published, the Agriculture Act set the scene for the future.

The Government of the day accepted much of the advice of the Wildlife Conservation Special Committee and, in 1949, established the Nature Conservancy to carry out official conservation activities. It got down to business with commendable speed. My first contact with it was in that year of 1949. Somebody in the Nature Conservancy knew that I had grown up on the edge of Romney Marsh; so I was asked to advise about the conservation implications of the War Department's holding at Dungeness. It was one of those few occasions in which wartime experience proved useful in peacetime. The range was used as an anti-tank gun range. I had trained originally as an anti-tank gunner and so I was well aware that at that time the projectiles of anti-tank guns were not explosive and that, despite their great velocity, they caused minimal damage to the environment. I also knew that the part of Dungeness outside the range had been destroyed by piecemeal holiday-shack development before the war, whereas the part inside it had remained virtually intact. Therefore, somewhat to the surprise of the Nature Conservancy, I pointed out that the range helped rather than hindered the conservation of this strange desert-like place.

The infant Nature Conservancy had much more difficult problems to tackle and certainly, when I joined its staff in 1953, it was too busy setting up its research stations and regional organisation to sit back and take a long hard look at what was happening in the British countryside.

I was appointed Regional Officer for South West England that year. Armed with a copy of Command 7122, I was told by Cyril Diver, the first Director of the Nature Conservancy, 'to get to know South West England better than anyone else'. In attempting to carry out his orders, I learnt at first hand that South West England contained a marvellous wealth of habitats. The war had affected it to some extent. Many woods had been clear felled and some had been replanted with conifers, many fields were better drained, some pastures had been improved, the great dunes of Braunton Burrows and much of Dartmoor and the Dorset heaths had been battered as military training areas. Even so, most habitats had survived into the mid-1950s without too much damage. The one exception was the habitat described at the beginning of this chapter – lowland heathland. It had little economic value as such and, once the Army stopped using it for military exercises and firing ranges, it began to be reclaimed extensively. This was obvious to a casual visitor, but since large areas still remained untouched there was little concern. People were aware of the reclamation in their own area but always assumed that there was plenty of heath left somewhere else, so there was no need to worry or to act. I was to encounter this attitude frequently on future occasions. For example, when working for the Commonwealth Scientific and Industrial Research Organisation (CSIRO) in Australia in 1972 I met many who did not realise that rain forests were threatened throughout that continent. There was an assumption that there was plenty more 'out there in the outback'. Of course most of the outback is much too dry to support rain forest. Somehow this was forgotten. The relatively small size of the area covered by rain forest was not realised nor the speed with which it was being felled (Fig. 10).

Conservationists cannot hope to protect threatened habitats unless they provide quantitative information on what is happening to the habitats and why it is happening. This was clearly the

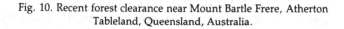

Fig. 10. Recent forest clearance near Mount Bartle Frere, Atherton Tableland, Queensland, Australia.

case in Dorset where the threat to heathland seemed greatest. So I set to work to discover how much had existed in the past, what had happened subsequently and the reasons for its destruction. All the heaths which remained had to be surveyed and the better areas identified so that they could be conserved.

Heathland is an unusual habitat. There is still so much of it in the British Isles that we tend to forget how special it is. Taking the world as a whole it is a very rare habitat, being virtually confined to the western fringe of Europe, the southern tip of Africa, and smaller areas in the southeast of Australia and a few mountain tops in central Africa. Heathland develops on poor, usually acid soils in areas with a temperate humid climate. In Britain the original heathlands were probably mainly coastal, but today the vast majority occur on areas which were once covered by woodland. In other words they result from human activity in the distant past. In Britain we are so used to a blend of 'natural' and 'man-made' that this does not raise any difficulties for conservationists. However, elsewhere it can. For example, in 1979 Professor R. Tomaselli told the Council of Europe's Group of Consultants on the Heathland Biogenetic Network about the problems he had encountered when trying to conserve one of the few heathlands which occur in Italy. Some of his countrymen affirmed that the heathlands were not worth conserving because they were not entirely 'natural'. Whether wholly or only partly natural, heathlands provide the only habitat for many species of plants and animals and hence most would agree that it is important to conserve them.

In Britain most heathlands are of the upland type. They are dominated by common heather or ling and support such species as Emperor Moths and Meadow Pipits. Like all heathlands they are liable to be afforested with conifers, but they are much less threatened than lowland heaths. This is partly because they were and still are so extensive and partly because one of their most characteristic species, the Red Grouse, has considerable economic value as a game bird and so its habitat is preserved. By contrast, the lowland heaths have almost no value today. In the past they provided poor grazing for cattle and ponies. The gorse which grows so prolifically on them was used for winter feed, bedding and fuel. Today these uses have almost disappeared save in the New Forest which still supports 3000 head of cattle and 2000 head of ponies.

The lowland heaths are not only more threatened than the upland ones but they also support a much wider range of species. Those in Dorset are especially rich. This is probably due to the fact that they occur on low-lying and sheltered land surrounding Poole Harbour and so are protected from extremes of weather. As a result they support species which are characteristic of warmer heathlands on the continent, notably the Dorset Heath (Fig. 11), the Sand Lizard, the Smooth Snake and the Dartford Warbler (Fig. 9). Two of the commonest and most characteristic species of lowland heaths in South West England, Bell heather and the grass called Bristle-leaved Bent, have very limited distributions in the world and are much more special than many realise. A heathland is like a miniature wood; under the heather bushes, leaf litter collects and provides a home for many invertebrate animals. Owing to the wetter climate of the South West the litter of its heaths is damper there and thus provides a much more suitable habitat than do drier heaths on the Brecklands and Sandlings of East Anglia. In the litter of

Fig. 11. Dorset Heath (*Erica ciliaris*), Hartland Moor, Dorset. This species is only found in Spain, Portugal, Western France and a few places in Dorset and Cornwall.

Dorset heathlands one finds two little cockroaches (Fig. 12) – the only wild members of a large family of insects to occur in Britain. Most cockroaches live in the litter of tropical rain forests, although two or three species have been able to colonise buildings and, until the advent of modern insecticides, were common pests in old houses in Britain and elsewhere. Another species commonly found in South West heaths and confined to them is the little bristle tail *Dilta occidentalis* (Fig. 12). This is a flightless insect of a group (the Thysanura) which has changed little since Devonian times when the land was first colonised from the sea. For these and many other reasons the heathlands are of great biological interest. When heather and gorse are in flower they are spectacularly beautiful. They have changed little for hundreds of years, except in extent, and thus they provide landscapes which we share with our forbears.

Originally most of the British heathlands were woodland – pollen analysis shows that hazel was an abundant species. Owing to the poverty of the soils it was easily cleared and early man did clear it. The land was overgrazed and the soil eventually became a podsol, that is one in which the basic salts are leached out, and in which a hard layer of iron salts often develops and prevents plant roots from reaching the nutrients beneath it. Only a small range of plants, notably heathers and some grasses, are preadapted to live in these rigorous acidic conditions. As a result they become the dominant species and, so long as colonis-

Fig. 12. A bristle tail (*Dilta occidentalis*) (a) and a cockroach (*Ectobius panzeri*) (b). Characteristic species of damp 'litter' under heather in Dorset (enlarged).

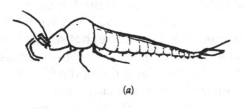

(a)

(b)

ing trees such as Birch and Scots Pine are kept at bay by fire or grazing, heathlands remain a remarkably stable habitat.

The first Ordnance Survey maps were produced when most of the poorer soils of lowland Britain were covered by heathland. Fortunately rough land was marked as such on the maps, and where it was shown on the poor soils which now bear heathland we can be sure that it was heathland then; although the amount of heather and gorse compared to grassland would have depended, as it does today, on the time and intensity of the last burn or the intensity of grazing pressure. The first Ordnance Survey map of Dorset, which was published in 1811, shows the situation as it was in the years just before the battle of Waterloo. During the Napoleonic Wars the need to grow more corn caused the reclamation of much chalk downland, but so poor were the soils of the Dorset heathlands that only small inroads were made on them. Practically all the land on the Bagshot Sands was heathland and it stretched from the outskirts of Dorchester to the New Forest in one great area, divided only by meadows on the alluvial soils in the valleys of the Stour, Piddle and Frome. Thomas Hardy was born on the edge of this great waste; in his novels he called it Egdon Heath. By 1896, when the Ordnance Survey map was revised, there had been some diminution and some fragmentation of the heath mainly due to building a new town where the little heathland river called the Bourne cuts its way through low sandy cliffs to join the sea in Poole Bay. By 1934 when the school children of England were collecting information for Dudley Stamp's Land Utilisation survey the Bournemouth–Poole conurbation had grown much more, yet the survey showed that 60% of the great Dorset heath present in 1811 still remained. I can recall the heath at that time as my family spent a summer holiday at Langton Matravers in 1933 and we visited Poole Harbour. I can remember the sensation of hot sun on sand and heath and the smell of salt water but my notes only recorded the gulls and the waders by the shore and the butterflies of the nearby chalk and limestone hills. It was many years before I visited the heath again and, by then, great changes were in the offing.

I spent 1946 training gunners on Salisbury Plain. With some-what ulterior motives I managed to organise an artillery exercise on the Dorset Heaths. I called it Exercise Hardy. We camped on

Stoborough Heath a few hundred yards from where I was to be working with the Nature Conservancy in a few years time. I rode my motorcycle to the top of Creech Barrow and looked at the great heathland which stretched from the army ranges to the sea near Studland. Later that summer I was able to explore these heaths on foot. Much had been used for military training and was deserted. The village of Arne (now in the centre of the Royal Society for the Protection of Birds – RSPB – reserve) was in ruins. It was very hot and desolate and Thomas Hardy's descriptions of Egdon Heath rang true. The absence of people and domestic animals gave it an eerie atmosphere of wilderness which I have never experienced in southern England in any other place except in the wilder parts of Dartmoor. By the time I was working at Furzebrook in the mid-1950s the heathlands of Studland had been divided from those of Hartland and Arne by vast conifer plantations at Rempstone. In my visits in 1946 I recorded a Peregrine on what was later to be the Studland National Nature Reserve, and a Hobby on Middlebere heath and a Redbacked Shrike by the Halfway Inn on lands which were to be later reclaimed for agriculture.

During the period 1953–60 when I was the Nature Conservation's Regional Officer for South West England, I was based at the newly established Furzebrook Research Station in the Isle of Purbeck. By the end of that period I had visited practically every heathland in what is now Dorset (but then was Dorset and that part of Hampshire lying west of the River Avon). Therefore I was in a position to compare the situation in 1960 with those of 1811, 1896 and 1934. When I plotted the changes since 1934 on a map they were dramatic (see Fig. 13). Forty-five per cent of the heathland existing in 1934 had been lost, and thus only a third of the original (1811) heath still existed. The original heath had consisted of about 16 areas but most occurred in three large blocks only separated from each other by river valleys. By 1960 the heath consisted of over 100 fragments.

The Dorset heaths have been studied intensively since 1960. Webb and Haskins (1980) showed that by 1978 only 6000 ha remained, that is 60% of what existed in 1960 and only 20% of the original heath. Fragmentation had increased further.

Despite the huge reduction of the Dorset heathlands remarkably few of their species have been lost so far. All or nearly all the

Fig. 13. Changes in the Dorset heathlands 1811 to 1960: (*a*) 1811, (*b*) 1896, (*c*) 1934, (*d*) 1960. (Moore 1962). (Crown copyright reserved.)

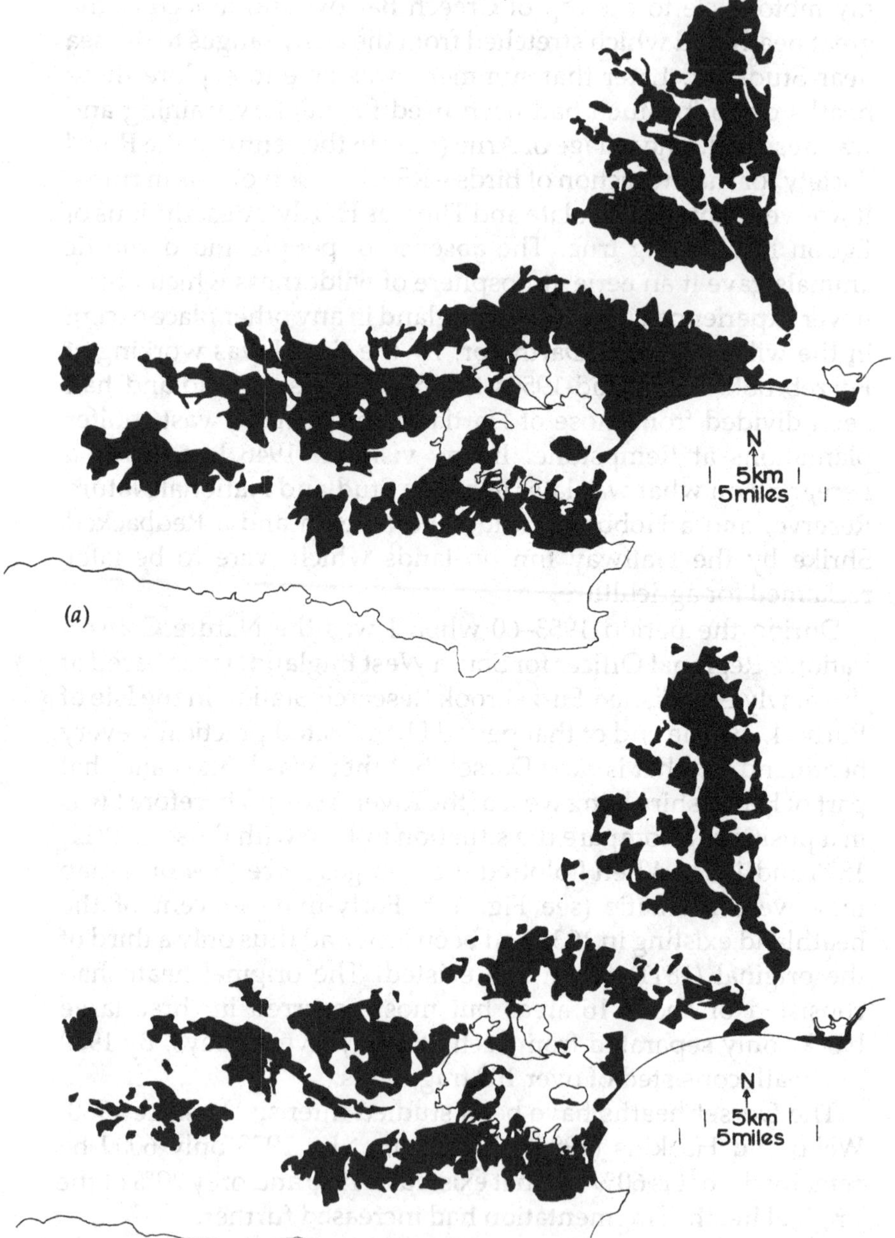

(*a*)

(*b*)

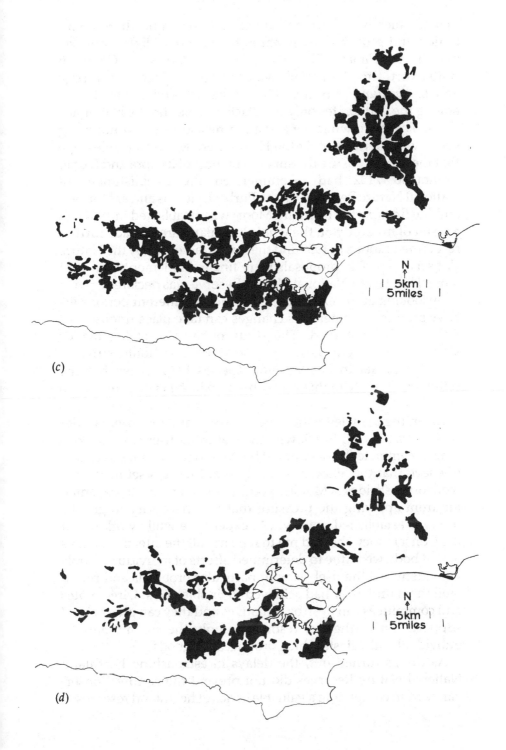

(c)

(d)

plants which have become extinct there had a northern distribution and probably disappeared because of a slight improve-
ment in the climate. They were Grass of Parnassus, Common
Butterwort, Crowberry and two Clubmosses (*Lycopodium selago*
and *L. clavatum*). No invertebrates have become extinct, and
among the vertebrates only the Black Grouse and the Natterjack
Toad. Therefore while conservation measures could not bring
back the wilderness of Egdon Heath there was reasonable hope
that they could ensure the survival of most of its flora and fauna.
Command 7122 had recommended the establishment of
National Nature Reserves on the Purbeck Heaths and at Morden,
and the first (part of Hartland Moor) was established in 1954. As
a result of my study on the whole heathland area I made further
recommendations in 1959 and these were accepted by the Nature
Conservancy. The proposals were finally implemented in 1984
when Holt Heath National Nature Reserve was declared.

By looking at the extent of heathlands at different periods we
have seen that momentous changes can take place unobserved
unless actively studied. The study of the Dorset heathlands
should have taught us that conservationists should strive to
identify threats to habitats and species before they become
self-evident. Only in this way can action be taken before it is too
late.

Given the knowledge that the Dorset heathlands were under
such threat it is fair to ask why the Nature Conservancy took so
long to complete the network of heathland reserves it has today.
I believe that the answer is simply that I left Dorset in 1960 to
work on pesticides at Monks Wood. My change of job prevented
me from applying the pressure that was necessary to get the
reserves established quickly. My experience while working for
the Conservancy showed me that nearly all the effective actions
of that body were due to determined efforts of individual people
who identified themselves with particular problems and promoted them until they had achieved their objectives. Directorates
and committees can destroy initiatives and they can give helpful
support, but in the last resort it is only the commitment of
individual enthusiasts which produces the goods.

As things turned out, the delays in establishing heathland
National Nature Reserves did not prevent the Nature Conservancy from completing a valuable series of heathland reserves in

Dorset which, with the help of the RSPB's reserve at Arne and the Dorset Naturalist Trust's reserves, should be able to conserve most of the denizens of Thomas Hardy's Egdon Heath and provide some links for them with the New Forest heaths to the east.

My study of the Dorset heathlands taught me a great deal. It showed me that most of a habitat could be destroyed within a relatively short period of time and yet most people – even those on the spot – failed to see what was happening. By measuring the decline of the area of heath, I was forced to take an historical view of the habitat; this led to my asking questions about its past use and why it had declined. It became starkly obvious that conservation was competing with powerful economic forces. In the Dorset heathlands these were primarily urban development, mineral extraction, state and private forestry, and highly sub-sidised agriculture. Only military training and the establishment of golf courses favoured the retention of some form of heathland (Fig. 14).

Fig. 14. The transformation of the Dorset heathlands; the situation in 1960 (Moore 1962). Most of the larger areas remaining in 1985 are now nature reserves. (Crown copyright reserved.)

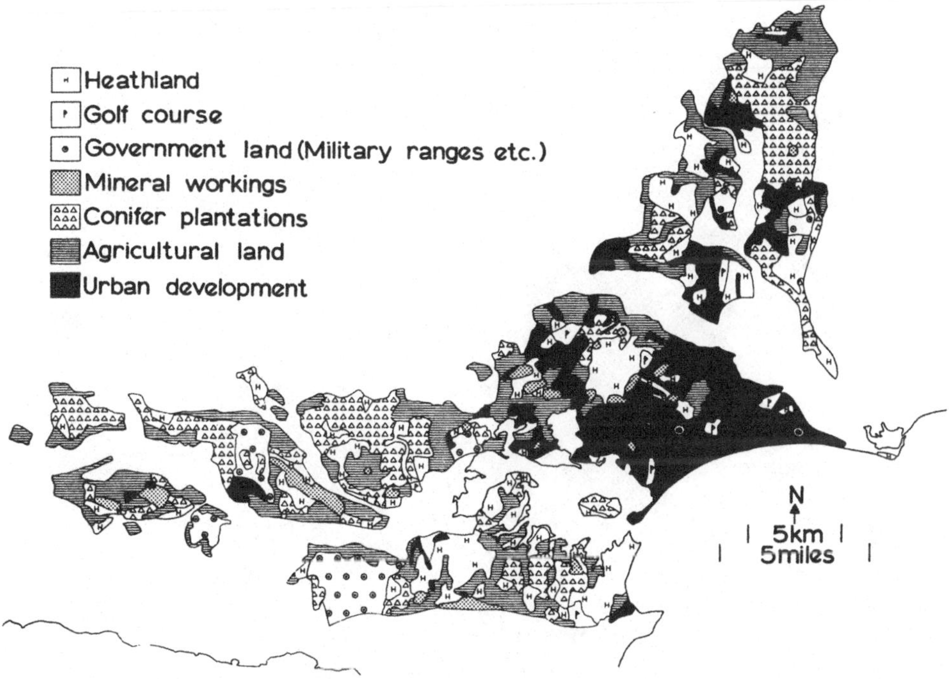

The economic forces which destroyed heathland had considerable political clout and so could only be resisted by individuals with a strong commitment to conservation who were armed with sound scientific information about heaths and about the effects of reducing habitats and dividing them into fragments.

My studies on the Dorset heaths were followed by others on other heaths in the East Anglian Breckland and in Surrey and Hampshire, and on other types of habitat in both England and Scotland. The rates differed, but all showed significant declines in the habitats on which most wildlife in Britain depends. Sadly, the case of the Dorset heathlands proved to be typical and not exceptional.

4

*

Disappearing hedges and ponds

When I first got to know Huntingdonshire in the early 1940s it was a closed, secret world of little grass fields surrounded by thick overgrown hedges. As most of the country was rather flat, one rarely got a view and so it was easy to get lost as one walked from one little green box to another with no landmark to guide one. It was a rather claustrophobic place. However, the lack of view was made good by the interest of the things at one's feet: the meadows were full of conspicuous plants like cowslips and inconspicuous ones like Adder's Tongue ferns. Today Huntingdonshire has a totally different atmosphere. It is an open land with wide views – one can see the shape of the low rolling hills. They are now covered by wheat and barley. A few hedges remain round the villages, which now appear exposed and island-like. The cowslips are confined to the verges of lanes and the sides of ditches, and the rarer grassland plants have become extremely rare or extinct. It is a good county to drive through but a dull one to walk in. People living in the district seem quite unaware of these huge changes. Many are too young to remember what it looked like before and, for others, the changes have been slow and gradual enough for them not to have noticed them.

Many things are not appreciated until they show signs of disappearing. This is as true of wildlife habitats and species as it is of people. It was particularly true of hedges. Hedges have been a part of the English scene for centuries, so much so that until recently they were taken almost entirely for granted. While many writers waxed lyrical over the English countryside, few stopped to think what made it distinctive. Hedges are found in most if not all countries where bushes can grow, but it is only in England and Wales (and, to a lesser extent, in Scotland and Ireland) that hedges *make* the lowland landscape. Yet they were largely ignored by writers, historians, archaeologists and biologists. When we began our studies on hedges in the early 1960s, the literature on hedges consisted of little more than a study of the role of hedges in maintaining spindle trees – because they provided the host plant of a pest, the Bean Aphid – and a study on our field boundaries by an ex-Luftwaffe pilot!

It shows how little biologists were concerned with the ordinary, that hedges received so little attention from them. Part of the reason was that hedges were known to be human artefacts, while it was still possible to believe that woods and heaths and downs were more or less natural and so more worthy of academic study. Today we are interested in studying the effect of man on nature both in the present and the past but this is a relatively new development. Up to the 1950s, human impacts on habitats and species were deliberately or subconsciously excluded from field studies. I suspect the reasons were complicated – a natural desire by biologists to simplify by excluding what seemed irrelevant, a snobbish distaste for the obvious and the practical by the intellectual, and an increasing loss of day-to-day experience of farming practice among most naturalists and biologists as Britain became more and more industrialised.

My own involvement with hedges began in 1960. I had been instructed by the Nature Conservancy to set up a Toxic Chemicals and Wildlife Section, which was to be based eventually at Monks Wood near Huntingdon.

My job was to organise research on the effects of pesticides on wildlife. For this we needed background studies on populations of wildlife in hedges and other farmland habitats which might be affected. As we have seen, extraordinarily little had been published on this subject. I can remember the exact moment when I realised that work on hedges must be an important part of our research programme – I was walking along the cliff path which goes from Studland to the Old Harry rocks at Handfast Point in Dorset. Beside the path was a hedge consisting largely of Field Maple (Fig. 15). This intrigued me as I had never seen a maple hedge before. It made me ask why hedges differ in their species composition. I realised that not only must we do work on hedges as the home of plants and animals which might be affected by pesticides, but that hedges were interesting in themselves and worthy of study for that reason too. I was sorry to leave the Dorset heaths but looked forward to studying the common hedge in eastern England.

Once based in East Anglia I began to record the different types of hedges which I saw there and elsewhere on my travels. In so doing I noticed that many hedges marked on the 1:25 000 Ordnance Survey maps had disappeared. Setting up the Toxic Chemicals and Wildlife Section did not leave much time for concentrated field work so I used train and car journeys to discover the scale of hedge loss. I found that it was occurring all over Britain and on such a scale that it must be having a major effect on wildlife. I published warnings in 1962 in the magazines *Countryside* and *British Birds*, estimating that between one-tenth and one-fifth of our hedges had already been destroyed in the last 20 years.

In 1963 I decided that our section should include a scientist with a knowledge of genetics. The post was duly advertised and Dr Max Hooper, a lecturer from Wye College, was the successful candidate. It soon became clear that he was also deeply

Fig. 15. Hedge, Studland, Dorset. This hedge is unusual in consisting largely of Field Maple (*Acer campestre*). It stimulated the author's interest in hedges in 1960.

interested in hedges and could make valuable contributions to our hedge research programme. By 1966, Max Hooper, Dr Brian Davis (the soil zoologist in the section) and I had done enough work to publish a summary of our reconnaissance studies on hedges. Among other things, we recorded the great variety of hedge types which we had observed, but we had no unifying idea to account for it. Shortly afterwards Max Hooper produced the hypothesis that the number of hedge shrub species was closely related to age – approximately, each species in a 30 yard stretch of hedge represented 100 years. Max Hooper was the first to emphasise that many exceptions could be found, but subsequent studies have confirmed the general validity of his hypothesis. It could be used by school children, local historians and conservationists and soon was. No other idea about hedges did more to get people interested in this habitat for which familiarity had previously bred contempt.

Meanwhile, my 1962 estimate of hedge destruction during the previous 20 years was getting increasing support from studies of aerial photographs made by Max Hooper and others (Fig. 16). In the absence of total coverage of aerial and field surveys there was much debate about the exact loss. Estimates made in the 1960s ranged from the Ministry of Agriculture's 2400 km (1500 miles) lost per year to the British Trust for Ornithology's 22 500 km (14 000 miles). Our own was 4800 km (3000 miles) per year. Controversy over the conflicting claims did much to bring the subject of hedges to the notice of the general public, while at the same time Hooper's Hedgerow Hypothesis and our studies on the flora and fauna of hedges made people more interested in hedges and so more concerned about their loss. At that time there were very approximately half a million miles of hedge in Britain, so that even if the Ministry of Agriculture Fisheries and Food's very conservative estimate of loss had been true, it still represented 0.3% loss per year which meant a very significant loss of wildlife habitat. In 1964 the little group at Monks Wood which was studying hedges was joined by Ernie Pollard. He started studying the insect fauna of hedges at Ashton Wold in Northamptonshire, and later obtained his doctorate on this study entitled 'Studies on the invertebrate fauna of hedges'. This was probably the first thesis ever to be based on the study of hedges, but it was not to be the last. By the early 1970s we had

Fig. 16. Loss of hedges in the parishes of Barham, Buckworth and Leighton Bromswold, Cambridgeshire (formerly Huntingdonshire) between 1946 and 1965. The dotted areas represent houses and gardens. (Pollard, Hooper & Moore, 1974).

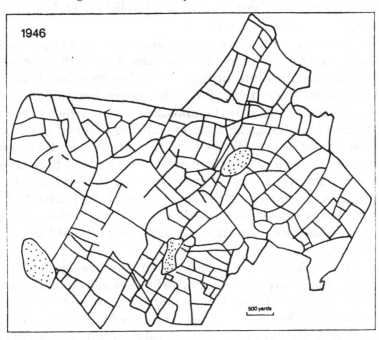

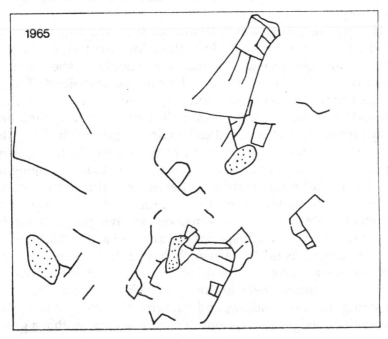

done enough work to provide material for a book devoted entirely to hedges, and in 1974 we published *Hedges* in the Collins New Naturalist Series.

By 1969 enough interest had been aroused for many people to 'want to do something about it'. Few questioned either the need for productive agriculture on the one hand or the value of hedges for the conservation of wildlife on the other; the problem was how to relate the two so that a landscape feature which owed its existence to traditional agriculture could be retained and managed so that it could still provide a habitat for wild plants and animals in the future. Clearly many hedges had still to be taken out in the interest of agriculture, therefore it mattered which would be retained and which could be demolished. Our studies in the Monks Wood district had also shown that the value of hedges as habitats for wildlife largely depended on how they were managed.

If anything effective was to be done about hedges, conservationists had to explain to farmers why they valued hedges and what it was that made one more valuable than another. At the same time they had to learn from farmers why farmers took out hedges, and why they used mechanical cutters to manage those that they retained.

Hedges are human artefacts and our flora and fauna evolved in their absence, so why does their loss matter? It matters because hedges provide habitats for woodland species in a country which has lost most of its original woodland. Today hedges make up about one fifth of the total area covered by broad-leaved trees and bushes in Britain. However, they are only second-best woods and so it is not surprising that they do not often provide a habitat for the rarer woodland species; they do, however, support large numbers of individuals belonging to the commoner more adaptable species, and this is their main value. For example, a mixed farm producing both crops and livestock, which has neither woods nor hedges, will rarely support more than eight species of birds, but a mixed farm with hedges around its fields will usually have at least twenty species and often may have forty or more, even if it contains no woodland. These figures refer to the farm itself and exclude species breeding in farm buildings and gardens. Put another way, the network of hedges throughout the country ensures that a great variety of common species remain part of the rural scene and can be enjoyed as part of ordinary life. We tend to take this for granted, but a walk across the new prairies of East Anglia shows all too clearly what lowland Britain would be like if all its hedges were to be removed.

Hedges are valuable as habitats for plant and animal species, and are also valuable in themselves. For centuries they have been a quintessential element in the landscape, particularly in lowland England and Wales. They emphasise land form in hilly country and give visual interest and structure in otherwise featureless flat country. Like farm buildings they add an historical dimension to the landscape because they were planted in different centuries; they outline county, parish and property boundaries and can even be used to trace the bounds of estates which have disappeared long since. Thus hedges support both the history and the natural history of the ordinary English countryside. Few hedges are so special that they have received special protection by law but, in aggregate, they make up an important element of the British heritage.

Little work has been done on the two-way relationship between the flora and fauna of woods and of hedges. In areas like the Weald and North Wales, where fields are small and hedges frequently contain trees, hedges seem to act as extensions of the woods so that one finds truly woodland species of birds, such as Nuthatch and Greater Spotted Woodpecker, nesting in trees in the hedge and feeding in both the hedges and in the small woods which they link. In such places it is easy for woodland species to disperse from one wood to another. The low, relatively young, and thus less varied hedges of the Midlands, of what Oliver Rackham calls the 'Planned Countryside', are much less like woodland but, for species which can only disperse from one wood to another via woodland, they must be crucial in maintaining links between woods. So, when these links are severed the woods become truly islands as far as these species are concerned. By recording the flora and fauna of isolated woods and plantations in areas which have never had hedges, such as the Cambridgeshire Fens, and by studying the dispersal behaviour of individual vertebrates and invertebrates, we have learnt that woodland birds, many insects and spiders disperse on a broad front and often at considerable heights and so do not require hedgerow corridors to travel along. Yet, even for these species, hedges must at times provide useful staging posts or stepping stones where they can feed and rest in between the wood they leave and the wood they will colonise.

The whole subject of dispersal between habitats has been sadly neglected by conservationists, and as the habitat islands become rarer and further apart much research needs to be done on it. However, at present, it appears that the conservation value of hedges as woodland habitats exceeds their value as pathways and corridors.

The value of a hedge as a woodland habitat depends to a large extent on how it is managed. Hedges without hedgerow trees are much less valuable than those with them, and hedges which provide no ground cover have very limited value for most wildlife. In the past the hips and haws of mature hedges provided a huge supply of food for visiting Fieldfares and Redwings from Scandinavia, but with the increase in growing winter cereals, hedge cutting is done much earlier in the year and, hence, hedges no longer provide this source of food in much of Britain today. This shows that we should be concerned not only with the loss of hedges, but with the kind of hedge which is allowed to survive in modern farming systems.

All these points about hedges have to be communicated by conservationists to farmers. Initially the farmer views hedges very differently because they affect his business and his livelihood. Hedges help or impede food production; they have to be managed and management costs money. Many farmers get much pleasure from their hedges but, unlike the passing walker or motorist, they have to do something about them and they have to make practical decisions about their upkeep.

To the farmer hedges have two main functions: they are fences to keep animals in and other animals (and people) out; and they provide shelter for crops and stock. In the past they were also an important source of timber and wood, and they always had a considerable role in game conservation.

Most hedges were laid down when all farming operations depended on manpower and horse power. With the invention of the internal combustion engine it became economically advantageous to use much larger machines, but small fields reduced their efficiency. Most hedges have been removed to enable farmers to use large machines as efficiently as possible. Other factors too have encouraged them to get rid of hedges: post and wire fences are easier to erect and maintain than hedges; they take up much less land, which can be an important consideration when land prices are as high as they are today. Hedges are mainly useful to farmers with livestock. In recent years many mixed farms have turned into purely arable ones. Arable farms not only require many fewer hedges but also the shade of hedges and hedgerow trees can reduce the yield of their crops. Even with mechanical aids, hedge management can be expensive. For all these reasons farmers often want to get rid of hedges and, with modern earth-moving equipment and draglines, it is much easier than it was to dig them out.

Conservationists have often said that farmers should conserve hedges because they reduce erosion and because they harbour populations of beneficial insects which help to control pests in the adjacent crops. There is little doubt that hedges must contribute to soil and crop protection to some extent, but the crucial question is, to what extent? I believe that in the present state of knowledge these two reasons for conserving hedges have been overplayed. Farmers are well aware that good crops can be grown and livestock reared in districts which never had hedges or from which hedges have been removed. At best the economic credit of hedges only outweighs the economic debit under certain circumstances or only in the very long term. The sad fact is that far too little research has been done to determine the exact value of hedges. We simply do not know the extent to which natural predators in hedges can reduce the need to use pesticides; we know very little about the significance of different densities of hedges in reducing soil erosion. Only governments can fund the long-term research which is necessary; so far, regrettably little has been done. Until more facts are available it will be extremely difficult to use the agricultural value of hedges as a convincing reason for their retention.

Enough has been said to show that farmers were right to get rid of many hedges in the interest of food production but that the conservationists were also right to be concerned about the effects on wildlife. The quarrel between conservationists and farmers about hedges has its roots in different perceptions and different values, and it has been obscured by the lack of knowledge about the economic and conservation significance of hedges. There is a clear need for better communication between farmers and conservationists, and for more research. These topics will be discussed in later chapters.

The other small-scale habitat which has declined greatly since the Second World War is the farm pond. Most ponds were dug to provide water for horses and cattle. They have disappeared because they are no longer required by stock. Large areas of eastern England, which used to support mixed farming, are now purely arable. There are virtually no working horses. Where cattle remain, clean piped water is now generally available, and this is preferred to pond water which can sometimes transmit diseases such as those caused by liver fluke.

When a pond is not required it is generally turned into a rubbish dump and eventually filled in, or it is fenced in and becomes surrounded by trees and bushes; it then becomes stagnant and eventually dries out. This process is hastened by improved drainage or water extraction, both of which lower the water table. As a result of all these changes thousands of ponds have disappeared. Two studies in Huntingdonshire and Leicestershire in the late 1960s showed losses of 35% and 30% respectively. No one has attempted to provide a national figure but there is not doubt that the losses have been very great. What are the consequences?

The farm pond is to natural meres and marsh what hedges are to woods: it is a human artefact provided by traditional agriculture which has provided a substitute habitat for many of the species found in the original natural habitat. Like the hedge, the pond has a disproportionate amount of edge and this favours marsh plants and the animals associated with them.

Long before man made and destroyed ponds, natural waterbodies appeared and disappeared quite rapidly. Ponds were formed by natural obstacles on low-lying land or were made by beavers. They quickly became colonised by waterweeds. They filled with silt and debris, were invaded by carr – that is, woodland adapted to damp conditions. Only plants and animals with good powers of dispersal were able to make use of these transient habitats. This ability to disperse and discover new wetland habitats before man dominated the landscape stood fresh water species in good stead when man added to what remained of the natural wetlands. It stands them in good stead today when man continues to make new water bodies such as clay and gravel pits and reservoirs.

The loss of ponds must have caused the decline of many aquatic species in many places, but few attempts have been made to quantify the losses. My colleague Dr Arnold Cooke has conducted surveys which show that the Common Frog has declined in many areas. Until recently all children living in the country were familiar with tadpoles. They could be caught easily and watched in the home or at school. When they disappeared many asked 'Where have the frogs gone?' Arnold Cooke showed that tadpoles are sensitive to certain pesticides which must have contributed to their decline. However, he concluded that the loss of their breeding grounds was probably the main cause of their disappearance.

Ponds take up much less space than hedges, and with modern digging machinery it is far easier to make a pond today than it was a hundred years ago. Therefore it is much more possible to compensate for the loss of ponds than the loss of hedges, and an increasing number of farmers do dig ponds for fish or wildfowl – or just to enjoy them. Many gardens are large enough to support a pond and many are made. Indeed Arnold Cooke has shown that in many areas the garden pond has been the salvation of the Common Frog, so that in much of England it is now commoner in towns than in the countryside. Better drainage can lead to acute water shortage in times of drought. This encourages farmers to make ponds for irrigation also. Increasingly these are designed to provide suitable habitats for fish and wildfowl as well as to provide water. All these developments help to make good the loss of cattle and horse ponds.

Ponds have always had a fascination for me. One of my earliest memories is of falling into one and then being put, clothes and all, in a bath and finding to my delight that I shared the bath with some tadpoles. Later I spent much time as a boy clearing bushes and trees from the edge of a pond in a wood in order to let the sun in. I was rewarded for my efforts: the pond was colonised by dragonflies including the very rare Scarce

Emerald Damselfly (*Lestes dryas*) (Fig. 17). Many years later in Dorset I used water-filled bomb holes to study territorial behaviour in dragonflies and its ecological consequences. By carrying out experiments in which I removed dragonflies or added dragonflies caught elsewhere, I was able to show that one could predict with considerable accuracy the largest number of mature male dragonflies that any pond could hold. These bomb hole ponds had been created when the Arne peninsula had been made a decoy in the war to divert attacks on military installations nearby. Their similar shape but variable sizes made them perfect for experimental purposes. By the time I worked on them the ponds had only existed for 9–14 years yet they supported stable populations of twelve species of dragonfly.

When I moved to the Fens to work at Monks Wood, we needed experimental ponds to study the effects of herbicides on aquatic organisms. So I got 20 of them dug in a field on the Wood Walton Fen National Nature Reserve. The field had been cultivated in the war and subsequently had become rough grassland

Fig. 17. Scarce Emerald Damselfly (*Lestes dryas*). A rare species found in ponds and ditches. It has greatly declined in England as the result of lower water tables.

with some thorn scrub. I have studied the dragonfly populations of these ponds ever since they were dug in 1961. By 1969 they had been visited by sixteen species of dragonfly and eight of these have maintained very constant populations on the ponds for the last 19 years.

When I retired from the Nature Conservancy Council in 1983 I received a very kind leaving present from my colleagues and ex-colleagues. I spent a part of it by getting a large pond dug in the field behind my house. A day and a half's work with a JCB and a skilled and enthusiastic operator produced a pond about 40 m long and 15 m wide (Fig. 18). The pond was dug in heavy impermeable clay in December. By March it was full of water. I stocked it with water plants throughout the summer of 1984. Many were given me by Colin and Joan Welch from their well-established pond in Hemington, Northamptonshire. By the end of the summer in 1984 no fewer than twelve species of dragonflies had visited the pond and I saw two others nearby.

Fig. 18. The author's pond, Swavesey, Cambridgeshire. This pond was designed to attract dragonflies. It was constructed in the winter of 1983. Fifteen of the thirty seven British breeding species had visited it by the end of the summer 1985. Five species bred success- fully during the first year of its existence.

Several species were seen laying eggs in the pond. Thus within one season the pond had been discovered by about a third of the British dragonfly fauna. Some had only to come from established habitats a few hundred yards away, but three species – the Emperor, the Broad bodied Chaser and the Large Red Damselfly – are rare insects in Cambridgeshire and must have come from much greater distances.

A pond is a small, but complete world, easy to take in at a glance and yet full of mystery beneath its surface. The great variety of creatures which colonise a new one do so with seemingly magical speed. The development of its flora and fauna is easily encompassed in a human lifespan. Ponds remind us of time and the interdependence of man and nature. Nevertheless the disappearance of ponds never became a major issue and cause of conflict as did the disappearance of hedges. One may speculate on why this was so.

Both hedges and ponds are widely distributed over agricultural Britain, although the total area covered by hedges greatly exceeds the total area covered by ponds. Both support a wide range of plants and animals whose original natural habitats – woods in the case of hedges, natural meres and marshes in the case of ponds – have been greatly reduced by man. The decline of both hedges and ponds mainly took place during the 30 years following the Second World War.

Like hedges, ponds vary greatly in age, but it is much less easy to calculate the age of a pond by noting the species it contains. So the historical interest of ponds is less obvious. I suspect that the main reason why the disappearance of ponds has caused relatively little concern is that they are much less conspicuous than hedges. Large-scale maps show how numerous ponds used to be but, even so, you often had to walk into a field to see whether it contained a pond. Therefore ponds have disappeared without many people noticing it.

It is also possible that when people have noticed the disappearance of ponds they have also been aware of the creation of new waterbodies: reservoirs, water-filled gravel and clay pits and newly dug garden ponds. They may have realised that the balance was at least partly redressed. In very general terms the loss of hedges probably does matter more than the loss of ponds.

The disappearance of hedges (and ponds) in the post-war years was not perceived as a major event until the early 1960s. The research on hedges at Monks Wood, which drew attention to the problem, was a by-product of a research programme on the effects of pesticides on wildlife. It took about 7 years of research and publishing scientific papers and articles to produce a national debate on hedges and hence on the role of hedges in modern farming and modern conservation. The historical significance of the debate was this: it forced everyone – not least the conservationists – to consider the value of the commonplace. It linked farmers and conservationists, first in argument and then increasingly in cooperation, and it got the general public involved with conservation.

5
*

The loss of special places

Many hedges and ponds could be lost without serious consequences for the country's flora and fauna because they were very numerous and widely distributed. The destruction of habitats which are rare and only found in a few places in Britain was a much more serious matter. I remember driving to the Lizard Peninsula in the far west of Cornwall 30 years ago, and seeing for the first time acres of rough grazing, pink with the flowers of the Cornish Heath (Fig. 19). One can only see this sight in a few places in western France, Spain and Portugal and in this one part of Cornwall. On the Lizard, Cornish Heath grows with a strange mixture of plants: some like the heathers are associated with acid soils, and others like the Carline Thistle, with alkaline ones. The soils in the area are very unusual. They are derived from the underlying ultrabasic rocks which are themselves very unusual. They include the beautiful serpentine rock, which is turned into those little mottled bowls and lighthouses which become souvenirs of thousands of Cornish summer holidays.

The agricultural value of the Lizard Heaths, like that of the Dorset heaths is very low, and so they too are much threatened by agriculture, forestry and development. They are obviously very special and need to be conserved as far as possible. The only practical way to do this is to turn them into nature reserves. Today 712 ha of the Lizard Heaths are protected in this way by the Nature Conservancy Council and a similar area is owned by the National Trust. Initially the Nature Conservancy found it difficult to establish a National Nature Reserve in the area and the process took a long time, so there was a considerable risk that heaths would be damaged or lost meanwhile. This type of problem was perceived by the founders of the Nature Conservancy and they included a special planning device in the 1949 National Parks and Access to the Countryside Act, in order to solve it and to extend conservation measures more widely. The device gave protection to places which were of special biological, geological or physiographical importance, but which were not being managed as National Nature Reserves. The Nature Conservancy was given the duty of selecting these places, which came to be known as Sites of Special Scientific Interest (or SSSI for short), and of notifying them to the local planning authorities. In turn, the latter had to consult with the Nature Conservancy whenever development requiring planning permission was proposed for the sites. Thus the device was only one of enforced consultation, but it prevented many sites from being destroyed through ignorance of their special interest; since planning authorities were loth to allow developments in SSSI, many sites were saved from unsuitable development. In

Fig. 19. Cornish Heath (*Erica vagans*). In Britain this species is confined to the heaths on the ultrabasic rocks of the Lizard Peninsula, Cornwall.

the years that followed many of them were turned into National Nature Reserves or into reserves managed by voluntary conservation organisations. More recently the national list of SSSI has developed a much greater significance than that of a partially protected waiting list of nature reserves.

The consultative procedure worked well when the threat to a site consisted of a development which required planning permission, but it did not cover other types of development. The regional staff of the Nature Conservancy, who were responsible for operating the SSSI procedure, soon found that much damage was being done to SSSI as the result of agriculture, forestry and other activities which were not subject to planning. For example, meadows were treated with herbicides; broad-leaf woodlands were converted into conifer plantations. In 1980, when the Secretary of State for the Environment proposed new conservation legislation, the Nature Conservancy Council and the voluntary conservation bodies saw it as a great opportunity to obtain better conservation for SSSI by bringing farming and forestry activities into the consultative procedure. However, many people who were not conservationists doubted the need to change the law because they were sceptical about the amount of damage that was being done to SSSI: it was clearly up to the Nature Conservancy Council to show conclusively that the damage was significant and hence that new legislation was required. It would have been a relatively easy task to demonstrate damage if all the records of it had been stored in a computer.

The necessity for using computers for analysing information in connection with National Nature Reserves and SSSIs had long been recognised by the Nature Conservancy Council but, unfortunately, the process of introducing a computer for site work had taken much longer than had been anticipated; in 1980 computer services were not available. The upshot was that I was asked to organise what could be done by other means in the very limited time available.

The aim of the exercise was to measure the extent to which SSSIs had been damaged during the year of 1980 and to discover the principal causes of the damage. From previous experience, we knew that damage could be anything from total destruction of the scientific interest of the site to insignificant local damage from which the site would quickly recover. Parliament would

only be concerned with significant damage. Therefore, significant damage had to be clearly defined and criteria laid down for
determining it. Significant damage to wildlife was equated with
what a farmer would consider significant damage if it had been
sustained by his crops or livestock. Thus damage was considered
significant if 1 ha or more had been destroyed or very seriously
damaged, or (in the case of a SSSI consisting of less than 1 ha), if
one-tenth of the site had been destroyed or very seriously
damaged. Damage was also considered significant if 10 ha (or
one quarter of small sites) had been affected by a deleterious
factor, such as drainage or spray drift, but had not been wholly
destroyed, or, if more than half the principal habitat of the site
had continued to suffer from a deleterious activity or from lack of
management necessary to maintain its scientific value. For example, damage was considered significant if a meadow notified for
its flora had not been grazed and was reverting to scrub and so
was losing its characteristic grassland species. Finally, if individuals of rare species especially protected by law had been
killed or removed from the site, this also counted as significant
damage.

The regional staff of the Nature Conservancy Council were
asked to report all damage to SSSIs which they had recorded
during 1980, and the criteria indicated above were used in
selecting those cases which could be listed as having sustained
significant damage. The information provided would sum-
marise what was known and would give some indication of
what was actually happening on the ground. However, the
figures obtained by this method would be an underestimate
because SSSIs covered a very large area – at that time about 5.5%
of the total area of Great Britain and so could not all be visited by
regional staff in one calendar year, let alone studied intensively.
In other words, significant damage could easily occur without
the Regional Officer or his assistants being aware of it. Accordingly, a separate sample survey was conducted in order to check
the information which had been obtained incidentally. Regional
staff were asked to select 15% of all the sites under their care
using a random method. They were then to make special visits
to these sites and report on any damage observed.

The results of the two surveys for sites notified for biological
reasons are summarised in Table 1. In practice it was not possible
to visit more than 13% of the sites; since the sites that were not

Table 1. Biological SSSI affected by significant loss and damage in Great Britain, 1980

Country	Routine Survey			Sample Survey				
	Number of sites	Number affected by loss or damage	Percentage affected by loss or damage	Number of sites in 15% sample	Number visited	Number affected by loss or damage	Percentage of sites visited affected by loss or damage	Percentage affected by loss or damage on the assumption that all unvisited in 15% sample sustained no loss or damage ± 95% confidence limits
England and Wales	2363	187	8%	337	313	46	15%	13.6 ± 3.7
Scotland	688	48	7%	106	86	13	15%	12.3 ± 6.4
Great Britain	3051	235	8%	443	399	59	15%	13.3 ± 3.2

visited tended to be remote ones and less likely to be damaged the final calculation was based on the assumption that the sites which had not been visited suffered no significant damage. Even when this assumption was made the results showed that damage to SSSIs was on a considerably larger scale than that demonstrated by incidental, routine recording.

Regional officers were also asked to record the causes of the damage suffered by SSSI when these were known. The results of the sample survey were likely to be the more accurate and are given in Table 2. As was to be expected, more than half the damage was due to agriculture. Of this, 15% was due to the cessation of traditional management; that is, due to the farmer not doing something beneficial rather than doing something immediately detrimental. Industrial damage included mineral working, tipping, laying pipelines, residential building and industrial pollution.

The use of fire is often necessary to maintain moorland habitats. The figure of 16% is confined to damage due to vandalism and badly controlled muirburn. Muirburn is essentially an agricultural practice. The damage caused by forestry included the conversion of broad-leaf woodland and moorland to conifer plantations. The 'other' category included harmful Water Authority activity, the parking of caravans, and severe damage to swards by vehicles. The study confirmed that the original procedure over planning developments worked reasonably well, but that there was an urgent need to obtain consultation over other activities.

Table 2. Causes of loss and damage to SSSI in 1980[a]

Agriculture	51%
Industry	21%
Fire	16%
Forestry	4%
Recreation	4%
Other	4%

[a]These figures refer only to SSSI where the main interest was biological.

The survey showed that in 1980 about 8700 ha of SSSI land suffered significant loss and damage. In about 2400 ha the scientific interest was totally lost. In isolation, 0.7% loss and damage to the total SSSI areas may seem unimportant, but it must be remembered that this figure represents only 1 year, and there was no evidence to suggest that 1980 was in any way exceptional. Also, the damage was being done to those places which were the most important for conservation in Britain.

The first results of the survey were made available to the House of Lords, in which the debate on the Wildlife and Countryside Bill had been inaugurated. The final results were made available for the House of Commons debate later in 1981. They were also given to the press and general public. Inevitably some distortion of the published facts occurred. For example, some protagonists claimed that in 1980, 13% of the SSSI had been destroyed instead of 'suffering significant damage'. Nevertheless, no one could deny that the damage to SSSI was significant, and much of it was due to activities outside planning control which, while being acceptable on ordinary farmland, were not acceptable on SSSI.

The survey also drew attention to the large amount of damage which was occurring through neglect, thus there was a good case for the nation to give financial support to farmers and landowners so that they could carry out positive conservation management.

The Wildlife and Countryside Act 1981 has had many critics, nevertheless a real attempt was made in it to deal with the problems which were highlighted by the 1980 survey of damage and loss of SSSI. Today the Nature Conservancy Council is forced by law to inform a landowner and tenant about any activity which might damage the scientific value of a site, and the landowner has to consult with the Nature Conservancy Council before he carries out any of the activities listed. By this arrangement, and by the continuation of consultation with the planning authorities, the Nature Conservancy Council should be fully consulted about practically all potentially damaging activities (vandalism can only be dealt with by the ordinary processes of law). For many years grants to help landowners manage SSSI for purposes of conservation have been available under the Countryside Act of 1968 and the Nature Conservancy Council Act of 1973, but the sums available were too small to have a significant effect. The new legislation greatly increases the scope for government support for conservation management on SSSI.

As with the particular studies on heaths and hedges, we can learn something from this study of damage to SSSI. First, there was yet further confirmation that a great deal of damage could occur without most people being aware of it. This was not surprising since the study of SSSIs in 1980 showed that few sites were totally lost. The damage was occurring unevenly but very extensively: the SSSI system was not liable to sudden collapse but to death by a thousand cuts.

The study also showed that damage was occurring to all habitats, not only those which had been studied specifically in the past. The scale of loss and damage drew attention to the magnitude of the measures which would be required to protect them in the future. If farmers and landowners were to be compensated for loss of opportunity on SSSI, the sums of money involved would be considerable. Whatever arrangements were made, it was becoming increasingly obvious that effective conservation could not be achieved painlessly and without expense.

The situation has been greatly improved by the Wildlife and Countryside Act but no one can predict how successful the new procedures will be. It is already obvious that the staff of the Nature Conservancy Council is far too small to carry out the enormous new duties of assessment and consultation. Delay in renotifying existing sites may result in loss and damage. It will certainly postpone the notification of new areas which should be notified and so they will be at risk. There was also a loophole in the law, which enabled an unscrupulous landowner to destroy the scientific interest of a newly proposed SSSI during the 3-month consultation period. For example, in 1983, the Nature Conservancy Council contacted the owner of one of the last remaining flower-rich meadows in Northamptonshire in order to notify the meadow as an SSSI; whereupon the owner threatened to plough it up. He carried out his threat before the 3-month consultation period, required by law, had elapsed. Similarly, a large firm of contractors forestalled the enlargement of the Torrs Warren SSSI in Dumfries and Galloway by bulldozing the sand dune ridges and filling in the hollows before the extension could be notified. All this means that it is essential to monitor what happens to SSSI as a routine measure.

The loss of wildlife habitat is now occurring so extensively in Britain as a whole that the need to conserve SSSI effectively is growing each year. Until recently SSSI had not been a household word, but a number of famous cases such as those of the Amberley Wild Brooks, Horton Common and West Sedgemoor, has drawn attention to the phrase, and more importantly to what SSSI are and why they matter. As a result their selection has become a matter of considerable concern and will be discussed in Chapter 7.

6

*

Choosing National Nature Reserves

The worst thing that can happen to a wild plant or animal is to lose its habitat: it simply cannot survive without it. Therefore the safeguard of habitats is easily the most important conservation activity. It underlies the establishment of nature reserves, the notification of SSSI and retaining hedges and ponds on the farm. Agriculture, forestry, industry, housing and recreation all make legitimate claims on the land, so obviously it is not possible to protect all the wildlife habitats which happen to exist today. Therefore conservation priorities have to be worked out and selections made. In doing this we are continually forced to consider the purposes of conservation and the biological and economic constraints on its practice. Therefore the ways we have selected sites in the past throw light on how ideas about conservation have developed. In this chapter I shall discuss the selection of National Nature Reserves, for which the beleaguered SSSI have provided a waiting list.

I know few pleasures greater than standing in a newly established nature reserve which one has helped to set up. You feel that you have made a special link with the plants and animals around you and with the people who will come to look at their descendants in the future. It is difficult to analyse why this is so satisfying. It is not just because a conservation battle has been won. I suspect that the underlying pleasure – relief is almost a better word – is connected with a desire to produce something permanent in today's world, in which so much else is subject to unpredictable change. Setting up a nature reserve has similarities with painting a picture or writing a scientific paper: a great deal of hard work has been done, obstacles have been overcome and, at the end, something new has been created. One has impinged on the future as well as the present. Of course, nothing can be predicted with absolute certainty – political madness or war may undo one's work or mismanagement may mar it – but what can be done has been done.

Establishing a nature reserve is a complicated business and frequently involves team work between biologists, land agents and administrators. All have a share in the satisfaction of achieving the final result, but in this chapter I shall only discuss the activity which has to precede every reserve acquisition – the initial part of the process – the selection of one place rather than another.

Today there is much debate on the function of nature reserves and SSSI and on the criteria for selecting them, but in the early days of the Nature Conservancy there was surprisingly little. In this chapter and the following one on SSSI I shall describe how I have seen ideas on this subject evolve.

As we have seen, the Nature Reserves Investigation Committee had made the fundamental proposal that government should become involved in nature conservation by setting up and administering National Parks and National Nature Reserves. Regional sub-committees had been established to list areas of special interest. The Nature Reserves Investigation Committee devoted much time on the philosophy of selecting nature reserves and in drawing up a list of reserves in accord with its philosophy. They sought 'a fair balance . . . between the conservation of representative types of flora and fauna and the protection of populations, aggregates of species, individual species and communities that are peculiar, rare or unique', they recognised the importance of primitive, i.e. ancient habitats, regional differences and transitional types of vegetation and the importance of having large reserves. Essentially, they selected the best-known examples of eight habitat types (coastal, fresh water, marshland, bogs and moorland, heathland, grasslands, woodlands and alpine habitat) but they were fully aware that these examples could not protect all individual species and types of vegetation, and so other reserves would have to be created to cater for them. The Committee were aware of many gaps in knowledge about the country's flora and fauna which increased the difficulties of selecting reserves. It says much for the wisdom of the Nature Reserves Investigation Committee that its deliberations could be used not only as a base for the government White Paper on the Conservation of Nature in England and Wales (Command 7122) in 1947, but also for the Nature Conservancy's Nature Conservation Review in 1977. In recent years, much work has been done on the refinements of reserve selection. This has proved a particularly exacting and difficult task but none of it has led to revising the basic tenets laid down by the Nature Reserves Investigation Committee in the middle of the Second World War.

Following the lead of the Nature Reserves Investigation Committee, the authors of the Conservation of Nature in England and Wales (Command 7122) made it clear that the purpose of National Nature Reserves was to conserve and maintain a resource and to provide areas for survey, research, experiment, education and amenity. They recognised that the requirements for experimental and educational reserves differed from those whose main role was the conservation of their flora and fauna as a resource. The difference between the purposes which they saw National Nature Reserves fulfilling and those which we see today was only one of emphasis. Nevertheless it is very important. It is made clear by their statement that their list of seventy-three proposed National Nature Reserves 'represents, in our considered opinion, the minimum necessary for a balanced representation of the national resources of wildlife in England and Wales, and therefore the smallest field in which the essential scientific work can be carried out to the national advantage'. The authors pointed out that the total area required was only 28 300 ha (70 000 acres) or 2% of the land surface of England and Wales. They saw the series of National Nature Reserves essentially as a mechanism for providing protected examples of all the main habitats found in the country. This was a reasonable view when most habitats were still fairly extensive and appeared not to be under serious threat.

However, as threats grew, the establishment of National Nature Reserves became more important as a means of protecting habitats as such. In other words, the conservation role of National Nature Reserves began to shift from the protection of examples to the protection of total resources. Generally this meant that more of each habitat had to be established as a National Nature Reserve. The situation is shown in diagrammatic form in Fig. 20. In a, much habitat remains and so the proportion which is National Nature Reserve (shown in black) is only a small proportion but, when the total amount of habitat becomes reduced, the proportion that is National Nature Reserve is much greater although its area has not changed (see b). Then, once the National Nature Reserve system develops a role of resource conservation it becomes necessary to increase the area and hence the proportion of the National Nature Reserve component of the total habitat is increased further (see c in Fig. 20).

Those responsible for drafting the National Parks and Access to the Countryside Act 1949 wisely stated that the function of National Nature Reserves was to provide both for research and for preservation. Thus the Act did not need to be revised when the role of National Nature Reserves changed as the result of habitat loss.

For many years the Nature Conservancy used the list of seventy-three sites provided in 'Conservation of Nature in England and Wales' and the list of twenty-two sites in 'Nature Reserves in Scotland' (Command 7814) as its shopping list for National Nature Reserves. The lists had been chosen so well, and the practical problems of conserving the sites were so great,

Fig. 20. The representation of a habitat by National Nature Reserves: a sequence of events that will increase. (*a*) The habitat is extensive, the National Nature Reserves are merely examples of it. (*b*) The habitat declines in area so the proportion that is made up of National Nature Reserves increases. (*c*) The area covered by National Nature Reserves is increased in order to protect what is left of the habitat, most of which is now nature reserve.

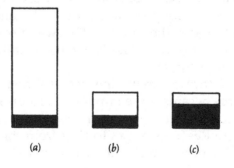

(*a*)   (*b*)   (*c*)

that the theoretical basis for site selection was not reassessed. This was not really surprising. Nevertheless, experience in establishing National Nature Reserves soon showed that the original shopping list, good as it was, was not entirely adequate.

In 1953 the Nature Conservancy divided England and Wales into seven regions for administrative purposes. In that year I was appointed as the first Regional Officer for South West England. In those days the region consisted of Herefordshire, Gloucestershire, Somerset, Dorset, Devon, Cornwall and the Isles of Scilly. In the following pages I shall draw on my experience in the South West to illustrate the problems which we faced when selecting National Nature Reserves in the early days of the Nature Conservancy. I shall use Yarner Wood in Devon and Bridgwater Bay in Somerset as examples of National Nature Reserves largely acquired to further ecological research, Rodney Stoke in Somerset to illustrate the problems of devising criteria for selecting a representative example of habitat, and the heathland reserves in Dorset to underline the importance of size and the spatial relationship between reserves of the same sort.

The acquisition of National Nature Reserves preceded the setting up of a regional organisation, and when I started work in August 1953 the region already contained Yarner Wood, a National Nature Reserve, which had not been listed in Command 7122. It had been acquired partly as 'a good representative of the general ecological conditions obtained in the sessile oakwoods* that characterize the valley slopes fringing the east side of Dartmoor', but mainly it was established for general studies and experimentation, thus fulfilling the requirement of Command 7122 to set up experimental reserves in addition to those protecting examples of habitat. The Moor House National Nature Reserve which consisted of 4000 ha (10 000 acres) of Pennine moorland was established at the same time, also for research purposes. The fact that two out of the first seven reserves to be declared by the Nature Conservancy in England were experimental reserves demonstrates the ethos of the time – the emphasis on research and lack of awareness of the threat to habitats which would soon dominate conservation thinking.

---

* A sessile oakwood is a wood in which the sessile or Durmast oak (*Quercus petraea*) is the dominant tree.

Moor House has continued as a research reserve, but when in the 1960s the Dartmoor oakwoods came under threat, the role of habitat protection largely superseded that of experimentation at Yarner. Nevertheless, the roles were complementary. Two of Yarner's most interesting species returned or appeared as the result of experimentation! First, the clearance of some oakwood to make way for an experiment on the establishment of mixed deciduous species caused the reappearance of the very rare Blue Lobelia (*Lobelia urens*) (Fig. 21). This plant is a Lusitanian species, that is one whose centre of distribution is in southwest Europe. In Britain it is confined to the South West of England. It had been recorded previously from Yarner, but had apparently disappeared until clearing allowed seed to germinate again. The

Fig. 21. Blue Lobelia (*Lobelia urens*). This rare plant reappeared on the Yarner Wood National Nature Reserve as an unsought by-product of an ecological experiment which disturbed the ground vegetation.

second species was the Pied Flycatcher (Fig. 22), a migrant bird which breeds commonly in Wales and Scotland and the north of England but was extremely rare in the South West. Dr Bruce Campbell carried out experiments at Yarner on the preferred nesting heights of hole-nesting birds, such as tits and redstarts, by putting up experimental nest boxes at different heights. Before he put up the boxes at Yarner, he predicted that the Pied Flycatcher would colonise the wood. He had studied the species in the Forest of Dean and was sure that Yarner would provide an excellent habitat for Pied Flycatchers if the absence of nest holes could be remedied. It was, and the Flycatchers turned up the first year of the experiment and have returned to the wood every year since.

Subsequent experience has reinforced my view that moderate physical disturbance of habitats in British nature reserves is nearly always beneficial. I suspect that the reason for this is as follows. Marine habitats are very stable, but land and fresh water ones have always been subjected to fairly violent events – lightning strikes, invasion by large mammals, wind-blow, flooding etc. In addition the normal cycles of birth and decay, including the death of large forest trees, produced much variation as well. Today most habitats in Britain exist as islands, and owing to their small size, tend to remain unaffected by such events for a long time, and therefore the species which depend on change

Fig. 22. Pied Flycatcher. This species colonised the Yarner Wood National Nature Reserve after nest boxes had been set up in order to study the nest height preference of redstarts and tits. Over 30 pairs now breed in the wood each year.

have fewer opportunities to exploit it. The situation which pertains in small woods today is abnormally static so, when a conservationist interferes by conducting an experiment, he imitates the pattern of the past in miniature by opening up the habitat, and some species can cash in on the disturbances he makes. The coppicing of woodland is the most significant example of interference. In the past, most woodlands in England were managed on a coppice and standards rotation. In other words the undergrowth was cut at intervals from 5 to 20 years to provide wood for poles, fencing etc., and the trees (at a much greater time interval) for timber. Coppicing produces a regular sequence from bare earth through scrub to high forest, which recapitulates what happened when areas were irregularly subjected to natural interference by fire, wind-blow etc. The experimental clearances at Yarner made openings in what had become an even aged oakwood. The past management of Yarner for oak bark for tanning did not allow the development of ancient trees with holes in them – the nest boxes simulated a habitat which had been an integral part of the ancient forest subjected to random natural accidents.

In 1954 another National Nature Reserve was established in South West England which, like Yarner, had not been listed in Command 7122. This was Bridgwater Bay. The Nature Conservancy has always taken a pragmatic view of reserve acquisition: it has responded to opportunity as well as to threat. Bridgwater Bay was one of the first responses to opportunity. The Somerset River Board had acquired over 6000 acres of the bay for coast-protection purposes and, on their own initiative, approached the Nature Conservancy about making it a National Nature Reserve. The Nature Conservancy readily agreed and, in 1954, this huge area of mudflats and salt marsh became a National Nature Reserve. Its main function was to provide an open-air laboratory for the study of coast erosion and coast defence. Much work was done there subsequently by Professor C. Kidson and Dr Derek Ranwell. The area was also very important for bird conservation, since it harboured the only moulting ground of Shelduck in British waters as well as providing feeding and roosting grounds for large populations of other wildfowl. If its great ornithological interest had been known when Command 7122 was being written, Bridgwater Bay might well have been included in the list of proposed National Nature Reserves. Thus Bridgwater Bay reminded us that primary survey of ecological interest in Britain was by no means complete; it was becoming clear that important new sites would be discovered and should be added to the shopping list.

Yarner Wood and Bridgwater Bay were experimental reserves and so were rather special cases, and they did not provoke any fundamental thinking about the selection of National Nature Reserves. Rodney Stoke, on the other hand, provided a great deal of food for thought, and certainly made me aware of the extraordinary difficulties in devising a rational and more exact system for choosing places to act as examples of habitat types. Its selection was so formative that I shall describe it in some detail.

Rodney Stoke is an ashwood (Fig. 23). Ashwoods depend on an oceanic climate, and are better represented in Britain than anywhere else in Europe. Therefore it is particularly important that the series of National Nature Reserves in Britain should include good examples. Ashwoods depend on calcareous soils as well as humidity, and so they are restricted to the limestone areas in the west and north of the country. Most of these are associated with deposits of carboniferous limestone, the rock which is such a feature of the Peak District and the limestone pavement country of the north Pennines. The only substantial exposure of this rock, and indeed of any limestone in South West

Fig. 23. Rodney Stoke National Nature Reserve, Mendip, Somerset. This was the first National Nature Reserve to be selected after a survey of all the examples of its ecological type, in this case Mendip ashwood.

England, occurs in the Mendips, and hence this is the only district in the South West which supports numerous ashwoods.

Command 7122 recognised that the Mendip-type of ashwood should be represented in the national series of reserves and it recommended Cheddar Wood as 'one of the few unspoilt examples of natural ashwood in Southern England that is not manifestly in a transient stage of development'.

Before joining the Nature Conservancy I worked and lived in Bristol and had got to know the Mendips and some of its woods. I was not entirely satisfied that Cheddar Wood was the best example of a Mendip wood, and I felt that before negotiations were started to establish it as a National Nature Reserve the final selection of the Mendip ashwood should be based on a survey of all the woods in the district. My proposal to make a survey was accepted by the Nature Conservancy and I was able to enlist the invaluable help of the late Mr Noel Y. Sandwith (the expert at Kew on the South American flora), who knew the Mendips well.

Noel Sandwith and I were well aware that the Mendips was an area of great botanical interest. Most of its outstanding treasures were found on rocks and rocky grasslands. The beautiful Ched-

Fig. 24. Blue Gromwell (*Lithospermum purpurocaeruleum*). A rare plant in England and Wales, it is characteristic of the Mendip Woods on Carboniferous Limestone.

dar Pink was found nowhere else in Britain. The White Rock Rose and the little umbellifer called Honewort were found only on the Mendip and on the isolated deposit of Devonian limestone at Berry Head. However, we knew that the woods on Mendip also harboured many other interesting species including populations of the Blue Gromwell (Fig. 24), a very local species with a southwesterly distribution, and of Autumn Crocuses. We knew that the Small-leaved Lime, a rather local tree in England and Wales, was a characteristic species of the Mendip Woods. In Cheddar it was dominant over much of the wood. We thought that our Mendip wood should contain these three species if it were to be truly representative. Calcareous rocks provide an especially suitable habitat for molluscs so we felt that the wood selected should also contain a good range of mollusc species. It should certainly have the Rough-mouthed Snail (*Pomatias elegans*) (Fig. 25) a species related to the winkle, and one of the two operculate snails to have become entirely terrestrial in Britain. If possible, the wood should contain the rare Bulin (*Ena montana*) (Fig. 25), which is a good indicator of ancient woodland. We started our search with these preconceptions.

About 200 Mendip woods were large enough to be marked on the 1 inch Ordnance Survey map. Of these, we discarded those which were largely on deposits other than Carboniferous limestone, those that were obviously recent plantations and those which were likely to be too small to support viable populations of plants and animals. In the end we looked at about ten woods.

Fig. 25. The Bulin (*Ena montana*) (a) and the Rough-mouthed Snail (*Pomatias elegans*) (b). The Bulin is an indicator of ancient woodland in Southern England. The Rough-mouthed Snail is confined to limestone districts but is much commoner than the Bulin. It is a close relation of the marine winkles (enlarged).

(a)   (b)

The brunt of the work fell on Noel Sandwith since it soon became clear that the botanical interest was paramount. The woods did contain interesting animal species but the more conspicuous species at any rate tended to be found in most of the woods we looked at.

When we had collected our data and had to decide which wood we were to select as a National Nature Reserve, I faced the problem that besets all conservationists when they have to choose the best example of a habitat from a long list. Our survey gave us a good idea of what the average Mendip wood was like today and a number of the woods would have been reasonable candidates, but did we really want a 'typical' wood in that sense? All the woods on Mendip had suffered a good deal of human interference over the years and so what was average today had been determined largely by human activities. Did what was typical in 1954 have any special significance? Instead, should we not select the wood that was most like the original forest which once covered the Mendip hills?

Of course, we did not know what it was like. Even by Roman times, the woods must have been modified since the Romans had carried out extensive mining operations in the area. However, in the early 1950s most ecologists underestimated the effects which man had had on vegetation over the centuries, and therefore we tended to think that existing habitats were more like the original ones than they were. As a result, choosing a wood like the ancient forest seemed easier than it was. Most of us in the Nature Conservancy at that time, as today, thought that National Nature Reserves should be as natural as possible, and so we preferred naturalness to typicalness (in the sense of approaching the average) as a criterion. We were encouraged in this belief by our empirical experience that the more natural habitats appeared to contain more species. Both Noel Sandwith and I were particularly interested in the conservation of species and therefore the criterion of diversity or species richness seemed especially important. Our studies showed that Asham Wood in the eastern end of the Mendip was far the richest in species. This was partly because of its unusually sheltered position but it was also due to atypical features – to the presence of underlying deposits other than Carboniferous Limestone and to a stream, a tributary of the Frome, which flowed through it.

While surveying the Mendip woods we soon realised that we should have to take account of factors other than scientific value. Limestone is a very valuable commodity, especially in districts where it is rare. It had been quarried in the Mendip for centuries. Very large quarries were active in 1954. Both Cheddar Wood and Asham Wood were owned by quarry companies, and in both woods vegetation was often covered by dust from blasting operations nearby. Part of Asham Wood had already been quarried. The Mendip hills had not yet been designated as an Area of Outstanding Natural Beauty and so there was no guarantee that even the most conspicuous woods would be protected from quarrying. We were well aware that we were dealing with a vulnerable habitat and its precariousness must also be taken into account. We also realised that very large sums of money for compensation would have to be paid if the Nature Conservancy bought a wood from a quarry company.

Clearly we had to relate all the conflicting values of the sites. It was tempting to develop a scoring system, but since there was no common denominator such a system could do no more than quantify our opinions. What seemed theoretically simple was in fact extremely difficult, and yet we had to come to a decision. As regards scientific value we felt that Asham was the best Mendip wood. It was easily the largest; it contained a wide range of limestone habitats as well as other ones, and it held many notable species; surprisingly it did not have the characteristic Blue Gromwell, but it did have Small-leaved Limes, Autumn Crocus and other notable plants such as Lily of the Valley, Solomon's Seal and Small Teazle. The local Wood White Butter fly occurred, as well as the Bulin snail *Ena montana*. In selecting Asham we gave priority to area and diversity, even though the wood as a whole was not, and could never have been typical of Mendip woods since it contained additional features not present in the other woods. At the end of the day economics took precedence over scientific appraisal. It became clear that the purchase of a large wood for which planning permission to quarry had already been granted would cost much more than the annual grant of the Nature Conservancy. Nor was an effective nature reserve agreement with the company possible, although attempts to obtain one were made. The Nature Conservancy had to settle for second best.

Cheddar Wood, the Command 7122 candidate was undoubtedly a fine wood. The abundance of Small Leafed lime in it was interesting but made it atypical. Much of the wood was very uniform. Like Asham it belonged to a quarry company. Rodney Stoke seemed marginally better than Cheddar. It was smaller but it had more variety of aspect and was unaffected by quarrying. Small-leaved Lime was present as well as good populations of Blue Gromwell and Autumn Crocus. There was a rich mollusc fauna; badgers had setts on the steep rocky slopes and buzzards bred in the wood and soared above its steep slopes. The Nature Conservancy accepted our recommendations and Rodney Stoke was declared a National Nature Reserve in 1954. When selling the wood to the Nature Conservancy, the previous owner found it convenient to include some small adjoining fields. These contained undisturbed grassland and scrub and supported several species of butterfly. As more and more grassland became ploughed up or was treated with herbicides, grassland of this kind became increasingly valuable, and today meadows are established as nature reserves in their own right. But the Nature Conservancy's first grassland was obtained as a matter of convenience to the seller!

In 1968 the Nature Conservancy responded to an opportunity and the wood of Ebbor Gorge, which belonged to the National Trust, was also declared a National Nature Reserve. It has much more oak than Rodney Stoke and other features which make it different from that reserve. It thus complements Rodney Stoke. Together they represent the ashwoods of the Mendips.

The establishment of Rodney Stoke National Nature Reserve had taught us many things. It revealed that the Command 7122 list of reserves, good as it was, was not an infallible guide to reserve acquisition: neither Asham nor Rodney Stoke was listed in it. It demonstrated that wherever possible the Nature Conservancy should carry out surveys of the whole habitat resource before selecting the best example, and it showed that the final selection of the best example was an extremely complicated business involving value judgements and economics as well as science.

My next foray into reserve selection demonstrated yet more complications. It taught me that not only was the size of a reserve important but that, once a habitat had been fragmented by man, the distance between the fragments had to be considered when setting up Nature Reserves. In Chapter 3, I described the rapid disappearance of heathland in Dorset and showed how this forced me to give priority to the conservation of this habitat. Command 7122 had recommended National Nature Reserves at Morden Bog and the Old Decoy Pond and on the heaths from Studland to Arne. Part of the latter, known as Hartland Moor, had been established as a National Nature Reserve in 1954 and additions to it were made in 1958. Morden Bog National Nature Reserve was set up in 1956 and additions to it were made in 1959. Were these reserves adequate considering the widespread destruction of heathland in the district?

When measuring the decline of Dorset heathland I carried out a concurrent study on the effects of the decline of heathland habitat on some indicator species of animal. My initial observations had suggested that the fragmentation of the heath had led to the disappearance of heathland species in the smaller fragments. I tested this hypothesis by recording the presence or absence of eight species in heaths of varying size and isolation. Four species – Dartford Warbler, Sand Lizard, Small Red Damsel fly and Silver Studded Blue Butterfly – were confined to heathlands and the four corresponding species used as a control were found in other habitats as well. They were the Stonechat, Common Lizard, Large Red Damselfly and Grayling Butterfly. It was expected that the last four species would be much less affected by isolation and fragmentation of heathland than the true heathland species and such was found to be the case (see Fig. 26). While all four of the heathland species still occurred in the large blocks of heathland none or only one or two species survived in the small outlying fragments, even though old records had shown that most had once been present in them. This study gave very strong support to the idea that fragmentation, and hence isolation, were detrimental to species; account should be taken of this when selecting National Nature Reserves. Since those days, a whole branch of biology known as Island Biogeography has been built up on this theme, and it is now clear that all islands, whether true geographical areas or habitat 'islands' have properties which are broadly consistent with each other and can be described mathematically; reasonable predictions about species number can therefore be made, if the size of the island and the degree of its isolation is known. This information was not available in the 1950s but my study on the Dorset heathland 'islands' did show that the size of nature reserves was more important than many believed at the time.

Size was only part of the problem. When Thomas Hardy was writing about Egdon Heath it was only separated from the New Forest by the flood plain of the River Avon. In the 1950s Poole and Bournemouth were rapidly expanding northward, and it was obvious that before long the heaths of Dorset would be separated from those of the New Forest by urban development. Populations of heathland species with poor powers of dispersal, especially those like the Sand Lizard which could not fly, would become increasingly isolated from each other. Where possible we had to consider providing stepping stones or staging posts between reserves to enable individuals to disperse from one area to another. Accordingly, in my paper to the Nature Conservancy

Fig. 26. The effects of isolation and fragmentation of habitat on eight heathland species in Dorset (Moore 1962). The numbers show the number of heathland indicator species observed on the heath indicated. The top figure in each circle shows the number of true heathland species which only occur on heaths, the bottom figure to the number of species which occur on other habitats as well. The figures suggest that the break up of the original heathland (see Fig. 13a) has led to a loss of true heathland species in the smaller more isolated fragments. (Crown copyright reserved.)

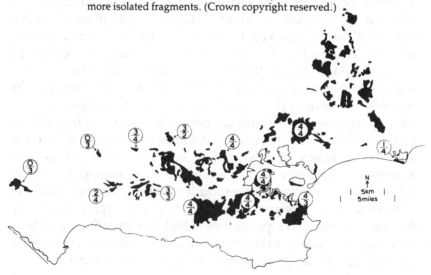

I made two recommendations: first, that the Dorset heathland reserves should be made as large as possible and secondly that there should be links between the Purbeck heaths and the New Forest. In practice this meant that Hartland Moor should be extended to include as much as possible of the surrounding heathland which still remained, and the National Nature Reserve at Studland which was under negotiation should be extended to include the neighbouring Godlingston Heath. It also meant that the Nature Conservancy should set up a National Nature Reserve at Holt Heath which, with the existing Morden Bog reserve and a proposed SSSI at Canford Heath, would maintain a link between Purbeck and the New Forest.

The work on the Dorset heaths emphasised the importance of area as a criterion for reserve selection and the need to relate reserves spatially so that at least some species could disperse from one to another. It also underlined the need to monitor changes in the total amount of each habitat so that nature reserves could be set up while there was yet time.

I do not want to give the impression that the problems encountered in South West England were confined to that region. They were not. By 1964, about half the 111 declared National Nature Reserves in Great Britain consisted of those proposed in the original White Papers Command 7122 and 7814. The remainder were not listed specifically in the White Papers, although many occurred within the Scientific Areas outlined in Appendix 8 of 7122 or within the 'special conservation area of North West Sutherland' described in 7814. Subsequent to the Dorset studies, loss and fragmentation of other habitats (notably other heaths and chalk grassland) were recorded. By 1965 the problems were widely discussed within the Nature Conservancy; the time was ripe for a new appraisal of reserve selection and my friend and colleague Dr Derek Ratcliffe was given the gigantic task of analysing the whole range of variation in the wild flora and fauna of Britain and of selecting a new list of potential National Nature Reserves in the light of the Nature Conservancy's first 16 years of experience. One of the most valuable features of this work was a critical analysis of the Nature Reserves Investigation Committee criteria on which National Nature Reserves had been selected hitherto, namely 'size, diversity, naturalness, rarity, fragility, typicalness, recorded history, position in an ecological/geographical unit, potential value and intrinsic appeal'. The case studies from South West England, which I have described in this chapter, show that all these criteria were already in use in selecting National Nature Reserves but, by listing them explicitly in 'A Nature Conservation Review' (see p. 89), and by discussing the relationship between one criterion with another Derek Ratcliffe did much to clarify the Nature Conservancy's attitude to reserve selection to the public at large.

The excellence of the original work of the Nature Reserves Investigation Committee ensured that official conservation in Great Britain was soundly based. Its very excellence may have delayed the reappraisal and confirmation of policy which was later enshrined in 'A Nature Conservation Review'. For 16 years the wisdom of the founding fathers of the Nature Reserves Investigation Committee seemed, and largely was, adequate for the Nature Conservancy's reserve acquisition programme. The determination of the individual members of the Nature Reserves Investigation Committee and the authors of the two White Papers on conservation ensured that the dreams in the war were put into action in peacetime. It was singularly fitting that the one member who was common to both the Nature Reserves Investigation Committee and the Wildlife Conservation Special Committee (who wrote Command 7122) became the first Director of the Nature Conservancy. He was Cyril Diver; we all owe a great debt to him.

7

*

Choosing Sites of Special Scientific Interest

The term SSSI conjures up an image of bureaucracy and officialdom. This is a very great pity for, as we have already seen, SSSI (together with the National Nature Reserves) are the most important places for wildlife in Britain. They deserve a better name than a stuttering acronym.

That something of such importance is known by a bureaucratic term reveals an unpalatable truth. People who care about nature tend to be romantics and prefer nature untamed. The very act of categorising the wilderness is repellent, especially if done by a government department. I can sympathise with this attitude, but it is totally unrealistic in the world of today. Nature is under serious threat and, if we are to conserve it effectively, we must solve the problems connected with it through the ordinary work day channels of finance, legislation and politics.

No conservationist can be effective unless one of his or her feet is planted firmly in the field and the other in the market place. This dual role gives the peculiar flavour to the work of the professional conservationist: part of his or her time is spent in some of the most interesting and beautiful places in the countryside, but the other part at meetings in large towns: gum boots and binoculars alternate with agenda and minute papers. The relevance of the paperwork to conservation is obvious to the practitioner, but it is extremely difficult to make administration an interesting subject for the reader. Yet, it is fundamental because the development of ideas about the selection of SSSI have reflected changes in attitudes about conservation, above all about its purposes and its relationship with other human activities and goals.

In Chapter 5, I showed that even though SSSI received some legal protection as the most important places for wildlife in Britain, many were being damaged and some destroyed altogether. This had a serious effect on the National Nature Reserve programme, because one of the main functions of the SSSI procedure was to give protection to places which would eventually become National Nature Reserves. In fact the first lists of SSSIs largely consisted of places recommended as National Nature Reserves in the government White Papers 7122 and 7814.

If SSSIs had never been more than potential National Nature Reserves, the method of selecting them would have been the same as that for National Nature Reserves, but from the earliest days of the Nature Conservancy they had a wider role and so the methods of selecting them were rather different. And, as changes in the countryside gathered momentum, our attitudes to conservation (and hence to SSSIs and their selection) evolved in response to the new situations.

The enactment of the Wildlife Conservation Special Committee's proposal on Sites of Special Scientific Interest (SSSI) was one of the most fundamental features of the The National Parks and Access to the Countryside Act (1949), although it did not appear so at the time. As we have seen in Chapter 5, the Nature Conservancy was given the formidable duty of selecting SSSIs and notifying them to the planning authorities. In a subsequent General Development Order, planning authorities were given the duty of consulting with the Nature Conservancy before giving planning permission for development on a SSSI.

Any area could be notified as a SSSI, so long as it was not being managed as a National Nature Reserve and so long as its scientific interest was special. No limits were put on the size or the number of sites. In the first instance, the Nature Conservancy notified the areas recommended as National Nature Reserves in Command 7122 and Command 7814, except for the few which had already been established as National Nature Reserves. Parts or the whole of the Scientific Areas listed in the appendices of those White Papers were also notified as SSSI. Much of the early years of the Nature Conservancy was spent selecting and notifying SSSI. As envisaged by its founders, an enormous amount of field work had to be done on 'individual assessment of the scientific claims of a large number of scattered sites'. Speed was of the essence and a remarkable crash course of action was initiated. In the Nature Conservancy the organisation of this immense task fell to the lot of the late Frank Green, an outstanding geographer and meteorologist whose wartime achievements in the Navy had included the cracking of an important enemy code. He was ably supported in the field by Dr Verona Conway, Dr A. S. Thomas and others. The fullest possible use was made of local natural history societies since there were only three Naturalist Trusts at that time. Today the naturalist trusts have largely taken over the role of natural history societies in the field of conservation. It should be remembered that, at that crucial juncture when it was vital to notify the bulk of SSSI in a very short time, it was the natural history societies which gave the small, inexperienced and understaffed Nature Conservancy the help that was required. Without the local knowledge of countless naturalists up and down the country, the task could never have been completed.

By the time the regional network of the Nature Conservancy had been established, most Regional Officers inherited very full lists of SSSI in the counties for which they were responsible. By the end of 1953, 1098 sites had been notified in England, 41 in Scotland and 21 in Wales.

No rigid criteria were given to those who selected the first SSSI. The law stipulated that the Nature Conservancy must notify 'any area of land, not being land for the time being managed as a nature reserve, which is of special interest by reason of its flora, fauna, geological or physiogeographical features'. In other words, the site had to have scientific interest and to be special. In practice, sites tended to be of two kinds: the localities of nationally or locally rare plants and animals, and representatives of particular habitats. Members of natural history societies were in a particularly good position to recommend the former. The latter were usually selected by Nature Conservancy staff or by experts like the late Mr Lousley working for the Nature Conservancy under contract. Everyone worked under great pressure against the clock. There was rarely time to survey sites very thoroughly. There is a story that Mr Lousley selected one SSSI in Cornwall from the window of a train! Those of us who knew Mr Lousley were not surprised when subsequent studies on the site showed that he had chosen well.

Looking back on this period I find the interesting thing is that those concerned found it relatively easy to select sites without guidelines or criteria, and that the net results were generally consistent. So, when in 1979 criteria were eventually laid down, nearly all the old sites notified still qualified under the new criteria. I suspect that the reason was as follows. All of those engaged in the work were good field naturalists. They had all looked at a vast number of places in the course of their lives. Their brains (their personal computers) had a great deal of information stored in them, and it was largely shared. Therefore, without much conscious effort they were able to select the special areas accurately and consistently.

After the initial crash programme, the rate of notification slowed down, but county schedules were revised as occasion demanded. These county schedules were lists of sites with maps and summaries of their scientific interest, which were sent to County Planning Officers, the Ministry of Agriculture, Fisheries and Food (MAFF), the Forestry Commission and others concerned.

In revisions of county schedules, newly discovered sites were added and those sites which had lost their scientific interest were denotified. As already mentioned, the loss of scientific interest was due to changes in agricultural or forestry practice in most cases. It was not always easy to make the right decision about damaged sites. I remember visiting a chalk grassland site in Dorset which had been reseeded, but it included so many Bronze Age barrows which had not been affected by ploughing that I decided to leave the site on the schedule. Now many years later archaeologists and biologists, recognising that they are often concerned about the same sites, are beginning to work together cooperatively. For example, there is considerable scope for combining archaeological and nature conservation on the numerous Iron Age forts which top so many hills in England and Wales. The steepness of their slopes has prevented ploughing, to the benefit of both archaeologist and naturalist. However, the interests of both are threatened if the earthworks are not grazed and, as a result, the slopes get covered by scrub. If the forts are grazed and not treated with fertilisers or herbicides they become joint monuments to man and nature. For example, the splendid ramparts of Maiden Castle in Dorset (Fig. 27) are grazed and still support populations of downland butterflies, though there is increasing evidence that sheep grazing rather than cattle grazing would usually be preferable from the points of view of both archaeology and nature conservation, because sheep produce a tighter sward and do less physical damage.

The Nature Conservation Review, which was so ably organised and edited by Derek Ratcliffe, represented the combined knowledge of the Nature Conservancy Staff and their successors, the staff of the Nature Conservancy Council and the Institute of Terrestrial Ecology. As we have seen, this impressive work was concerned with the description of British habitats and the selection of National Nature Reserves. In the selection of sites for National Nature Reserves candidate areas were divided into four grades. Grade 1 sites were those of international or national importance, all of which merited National Nature Reserve status. Grade 2 sites were equivalent or only slightly

Fig. 27. Maiden Castle, Dorchester, Dorset. The steepness of the ramparts has ensured the survival of the chalk flora and fauna. Grazing maintains both the archaeological and natural history interest of the site.

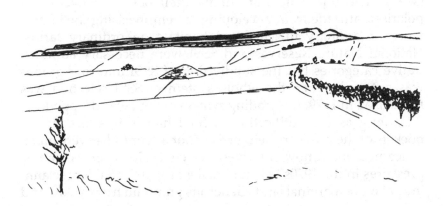

inferior sites which could be regarded as alternatives to Grade 1 sites if it proved impossible to safeguard the latter. Grade 3 sites were those of high regional importance, of SSSI but not National Nature Reserve quality, while Grade 4 sites were of lower regional importance. They 'would not qualify if they were located in a region with more extensive representation of the particular ecosystem'. They also were considered to be of SSSI quality.

Thus an SSSI could be a Grade 1 or Grade 2 site, awaiting establishment as a National Nature Reserve, or a Grade 3 or a Grade 4 site which was unlikely to be given National Nature Reserve status. Internally the classification was useful in helping the Nature Conservancy Council to order its priorities when dealing with a particular potential nature reserve. In the outside world it had the undesirable effect of giving some SSSI second class citizenship and thus reducing their chances of survival. Inevitably, County Planning Officers were more willing to allow development on a site that was labelled Grade 4. The same basic problem arises whenever land is classified. The very act of notifying SSSI at all makes it less easy to conserve wildlife habitats which are unnotified. Despite clear statements in the Wildlife and Countryside Act to the contrary, landowners and their agricultural advisers have frequently taken the line 'there is no need to take conservation into account in this particular place because it is not an SSSI'. There are signs that this attitude is becoming less common but, if it increased, the Nature Conservancy Council would have little option but greatly to increase the number of SSSI. The Farming and Wildlife Advisory Groups (see p. 104) and others are doing their best to prevent this polarised attitude from developing, by emphasising the importance of ordinary unscheduled habitats on ordinary farms. National Nature Reserves and SSSI were necessary administrative categories but the subdivision of SSSI into four grades was not and, when the whole system of SSSI selection was formalised in 1979, the grading system of SSSI was dropped.

So long as the notification of SSSI had little effect on the pockets of their owners neither they nor anyone else were much concerned about how sites were selected. However, economic pressures in the 1970s began to make farmers consider reclaiming yet more marginal land. Generous grant-aid from MAFF and the Department of Agriculture for Scotland (DAFS) and advances in mechanical and chemical technology made reclamation increasingly feasible. When proposals were made to reclaim SSSI, they had to be opposed by the Nature Conservancy Council. As a result, farmers increasingly disliked their land being notified, and increasingly they asked 'why should my meadow be notified and not my neighbour's?' Other owners began to ask the opposite question 'why shouldn't my land be an SSSI when my neighbour's is?' They did so because exemption from Capital Transfer Tax could be claimed under the Finance Acts of 1975 and 1976 if reasonable steps were taken to maintain the scientific value of an SSSI. The final decision in these cases rests with the Treasury; nevertheless, to date, all claimants for exemption from Capital Transfer Tax for land notified as SSSI have been successful. Thus the circumstances of the mid-1970s made it both more disadvantageous and more advantageous to own SSSI; which it was depended on the circumstances of each site. Generally speaking, the disadvantages were greatest on small sites in the lowlands especially if they were on relatively good land, and the advantages greatest on large sites. By 1975, the Nature Conservancy Council realised that the public's attention would be focussed increasingly on SSSI and so the haphazard method of site selection, however effective in the past, would be unacceptable in the future. Clearly a review of SSSI procedures was required and this was initiated in 1975. Two working groups were set up. The first was on fiscal incentives and was chaired by Miss Nettie Bonnar. Its job was to look at the implications of the Finance Acts on the Nature Conservancy Council's notification procedures. The second was on scientific definitions and criteria and was chaired by myself.

Circumstances had changed greatly since the passage of the National Parks and Access to the Countryside Act of 1949. As we have seen, threats to wildlife habitats were increasing and it was becoming increasingly difficult to conserve wildlife habitats on the ordinary farm except for those which supported common and adaptable species. This meant that the relative importance of SSSI was growing and so, quite apart from public relations aspects, it was crucial that the Nature Conservancy Council should select the SSSI with even greater care than hitherto.

We had to go right back to square one and ask fundamental

questions about the role of SSSI under the new circumstances. Wisely, the original legislation had not defined their exact purpose and so the Nature Conservancy Council could legitimately ask radical questions and propose radical solutions. We could not look at SSSI in isolation. Since many were key sites – that is, they were considered to be of such international or national importance that they should be established as National Nature Reserves or given comparable protection – there was a close link with the programme which selected nature reserves. Yet equally important was the link with the ordinary countryside itself which was not protected as a reserve or notified as a site. In the last resort, the number and nature of SSSI must depend on what is happening to this land (for which the term 'the wider countryside' is often used). So the working group on scientific definitions and criteria had to take a comprehensive view. We had to take cognisance of what was happening to the whole of Britain's wildlife resource, habitats and species alike. We had to measure the extent to which the existing system protected the resource and to identify its deficiencies. We had to develop a rationale for SSSI selection which was consistent with the selection of National Nature Reserves and would be broadly acceptable to the scientific and general public. Finally, we had to draw up very specific guidelines to enable our colleagues in the regions to revise existing schedules of SSSI and select new sites.

We knew that existing information on SSSI was inadequate, and that research must be undertaken to elucidate the problems which we knew about already, and those which were bound to arise as the work proceeded. A series of studies were made between 1976 and 1979 and will be described below.

As mentioned earlier in this chapter, most SSSI were chosen because they were the localities of rare species or because they were representative of a particular habitat. Each site was chosen on its own merits and, in the early days of the Nature Conservancy, no one thought of the totality of sites having a special role in British conservation. This was because most habitats still had many representatives – or were thought to have. By the mid-1970s we had enough facts about habitat loss to realise that not only species but whole habitats were threatened with extinction. There seemed to be little threat to the common adaptable species; the loss of hedges, grassland and ponds would to some extent be compensated by scrub on derelict land, gardens, motorway verges and gravel pits respectively. The last type of habitat was increasing so much that species which could live in it were likely to increase. Birds such as Great Crested Grebes, Tufted Duck and Little Ringed Plover were now common in areas such as Cambridgeshire where only 40 years ago they were rare or absent. Similarly the Common Blue Damselfly, the Black tailed Skimmer and the Emperor Dragonflies which quickly colonise gravel pits must now be far more abundant in southern England than at any time in the past. Nevertheless, most species are not adaptable and common but have exacting requirements and are rare. They are dependent on stable habitats which have been in existence for long periods of time, places like chalk grassland, heathland and ancient deciduous woodland – just the places which our studies on habitat loss showed we were losing throughout the country. It was clear that SSSI should be used increasingly as a means of conserving the best examples of what remained in order to ensure the survival of the species in the country as a whole. This led us to conclude that the totality of SSSI should be considered and the new function of the system should be to conserve the absolute minimum necessary to maintain our flora and fauna.

This concept was and, I believe, still is a useful one because it roughly defines the role of the SSSI system under present conditions. However, it cannot give practical guidance about selecting particular sites because we do not know, and are most unlikely ever to know, exactly what minimum is necessary to ensure survival. We know that some species have had a very restricted range in Britain for hundreds of years and yet have survived, for example the Cheddar Pink and the White Rock Rose mentioned earlier. Some, such as *Diapensia lapponica*, survive on a single mountain. Not all such species are plants; the Swallowtail, the Black Hairstreak, the Heath Fritillary and Glanville Fritillary butterflies appear always to have had very restricted ranges yet they have survived. On the other hand, other species with very limited ranges have become extinct such as the Alpine Butterwort, the Mazarine Blue and the Black veined White butterflies and the Norfolk Damselfly and the Orange Spotted Emerald Dragonfly. We are not dealing with absolutes, yet the destruction of one locality of a species which is confined to very few places is obviously more detrimental for that species than the destruction of one locality of a species which is found in many places. Therefore a large proportion of the localities of rare species should be notified and, for the very rare ones, all existing sites should be notified; in theory it can be shown that a number of medium-sized sites are likely to conserve more species than one very large site. But, in Britain, most habitat fragments are so small that only the larger ones can support populations of those species of birds and mammals which live at low population densities. Therefore, in practice, the general rule for size of site has to be the larger the better. All too often, habitats of the desirable size no longer exist.

As noted earlier, plants and animals differ greatly in their ability to disperse and the distance between two sites may be sufficiently small for one species to move from one to another, but for another species the distance may be too great. As shown in Chapter 2, sites would have to be linked by woodland or at least by thick hedges if we wanted to ensure that Black Hairstreak butterflies could move between their woodland sites. On the other hand, small birds such as the Goldcrest and Long Tailed Tits occur in small isolated woods in the Fens where there have never been connecting hedges, thus for them sites can be quite widely spaced.

As in so many conservation problems, no completely suitable course of action is possible and we can do no more than seek the best compromise. In practice, how far apart should sites be? Something could be learnt about dispersal from observing the rate of colonisation or recolonisation of suitable habitats. The very cold spring of 1963 exterminated the Dartford Warbler in the heaths of Surrey and north Hampshire but a few pairs survived in the Studland National Nature Reserve and elsewhere in Dorset and the New Forest. By 1968 they had begun to reappear in their more northern locality about 48 km (30 miles) away and, by 1974, they were breeding regularly there.

Until about 70 years ago, peat was dug in the Fens and this left small acid pools at fens such as those at Wicken and Chippenham. Those supported acid-water species of dragonflies, the Small Red Damselfly, the Black Darter, the Keeled Skimmer and the Common Hawker. When peat digging ceased, the ponds filled in and the acid-water dragonflies became extinct in the

Fens. Later I was able to test the ability of these species to recolonise the Fens by providing acid-water pools at the Wood Walton Fen National Nature Reserve (Fig. 28) where none had existed for many years. This reserve contains an interesting area known as the Heath Field which supports Sweet Gale, heather and other plants requiring acid conditions. With the help of the late Mr Gordon Mason, who was the warden at the time, we dug three ponds which soon filled with fairly acid water of pH 5.5. One pond was dug in 1974, the other two in 1977. By 1980, sixteen species of dragonflies had been recorded on the ponds. One of them was the Common Hawker, one of the lost acid-water species which had also colonised the new mere at the Holme Fen National Nature Reserve 6 km (4 miles) away. Its nearest recent locality was in the Norfolk Breckland 68 km (42 miles) from both Holme and Wood Walton. The Common

Fig. 28. One of three acid-water ponds dug at Wood Walton Fen National Nature Reserve to test the dispersal ability of dragonflies dependent on acid-water. They are the only acid pools for many miles around. So far only the largest of the acid-water species has discovered the site. Evidence of this kind was used in determining how many SSSI there should be in a given area.

Hawker is easily the largest of the lost acid-water Fen species and still occurs in several localities in Norfolk. The other three species are much rarer. The nearest site of the Small Red Damselfly is 84 km (52 miles) away, that of the Keeled Skimmer 100 km (62 miles) away, and that of the Black Darter 63 km (39 miles) away. All the populations are small and so it is not surprising that individuals dispersing from them have not yet discovered the little ponds in Wood Walton Fen.

We took this type of information into account when deciding on the 'mesh' of the SSSI network. We felt that if we selected the best example of each type of habitat within areas of approximately 250 000 ha (that is, 50 km × 50 km or 31 miles × 31 miles) this would enable numerous species to move from site. This area approximates to the size of a small English county and, for administrative reasons, it was much easier to use counties or districts as the areas of search for the best example than areas defined by map grid lines.

However sites are selected, there are bound to be anomalies because habitats are not evenly dispersed. Some areas or counties have many more examples of a given habitat than others, so that the quality of the best site in each area or county is bound to vary between areas. Common sense may decide that the best example in one area is not worthy of notification, whereas in another two examples should be selected. However, the value of a network of roughly evenly spaced sites is very great, not only because it allows for at least some dispersal between sites and hence genetic exchange, but also because it caters for changes in the range of species (which will undoubtedly occur as climate changes). For example, if winters became milder, species like the Dartford Warbler which is now confined to heaths in the south of England would be able to colonise South Wales and the Midlands – presuming heathlands still existed there. Therefore it is desirable to have heathland represented in as many areas as possible. If the climate became colder species like the Dartford Warbler might become extinct in Britain but a network of heaths would enable heathland species now confined to Scotland to move south.

Having decided that the notification of SSSI should be based on selecting the best representative of habitats in areas approximating to 250 000 ha, we had to decide exactly on what should constitute a habitat. The number of sites selected would depend on the number of habitat types. For example, if we merely selected the best broad-leaf wood in each area each county would only have one woodland SSSI and we would get very inadequate coverage (for, in one county it might be an oakwood and in another an ashwood). On the other hand, if we subdivided woodland into numerous subtypes the Nature Conservancy Council would find itself notifying an unacceptably large number of sites with relatively little additional conservation advantage. At the time of writing, this problem has not been entirely resolved because as yet there is no universally accepted system of habitat classification. Meanwhile Nature Conservancy Council officers are provided with suggestions on how far they should divide habitats (like woodland, heathland etc.) when making their search for the best representative example. For example, nine types of heathland are recognised in Britain, although most counties will only have examples of two or three types.

The selection of the best example of each habitat type within a county (the area of search) encounters exactly the same problems as the selection of National Nature Reserves, which were discussed in the last chapter. Indeed the best example chosen on the national scale would obviously be the best example in the smaller area in which it occurred. All the objective criteria used for selecting National Nature Reserves would also have to be used in selecting SSSI. Very small sites should be discarded and naturalness and diversity should be the main criteria used.

The SSSI system has to embrace the National Nature Reserves series but also to be firmly related to the conservation of the total resource. In practice, this means that the number of examples of each habitat within an area of search has to be related to the scarcity of the resource and its vulnerability. Therefore we designed a sliding scale by which the rarest and most vulnerable habitats received the greatest protection and the commonest and least vulnerable the least. For example, all examples of habitats whose total area in Britain is less than 10 000 ha have to be notified as SSSI if the habitat has been reduced by more than 10% since 1945. Whereas more than one (but normally not more than 5 sites) have to be notified if a similarly reduced habitat's total area exceeds 10 000 ha.

The conservation of habitats is an excellent method of conserving individual species. For example, no National Nature Reserve has been established especially to conserve dragonflies, yet thirty-two species of dragonflies (84% of the species occurring in Britain) are to be found in National Nature Reserves. This is because there is so much variety in the physical and chemical characteristics of the waterbodies in the National Nature Reserve series, and so great a geographical range, that suitable habitats are provided for nearly all the British species. Nevertheless, quite a number of species, notably of mammals and birds, are not so obliging and depend on localities which would certainly not be chosen as the best example of a habitat. For example, the roof of Furzebrook Research Station near Wareham in Dorset, the building which housed my Regional Office in the 1950s, was the home of two of Britain's rarest bats. The bats had chosen well since Dr Bob Stebbings, an international expert on bats, also worked there and it was he who discovered them. Not far away, another rare species (the Greater Horseshoe Bat) had one of its main roosts in a ruined outhouse of Bryanston School. So important was this roost for the species that the National Heritage Fund provided a large sum of money to enable the ruined building to continue as a roost for this bat.

British estuaries and their surrounding salt marshes provide passage and wintering grounds for thousands of waders, duck and geese which breed in other countries. Some of these marshes, for example those which fringe the North Norfolk coast, also have great botanical and physiographic interest; others do not, yet clearly they should be protected for wildfowl alone.

The localities of some of our rarest plants and invertebrates would not be chosen as outstanding habitats. For example, one of the main localities of the wild Grape Hyacinth is a hedge bottom by an otherwise unremarkable roadside verge in Cambridgeshire.

Therefore, additional criteria had to be defined so that localities of the types described above could also be notified as SSSI. For example, any sea bird colonies which contains 1% or more of the total British breeding populations of a species is notified as a SSSI. Of course, this does not mean that we know that the survival of the species depends on the colonies protected being 1% rather than 0.5% or 5%, but the figure is probably approximately right and it ensures consistent treatment throughout Britain. It can always be modified in the light of new knowledge.

Before we could recommend our guidelines for selecting SSSI to the Nature Conservancy Council we had to undertake a number of special studies to see whether they were practicable. We had to steer a middle course between selecting too few sites to maintain the resource and too many which would result in the coinage being debased: SSSI must be kept special. For example, we felt that it was theoretically desirable to give protection to all localities of plants listed in the Red Data Book – that is, plants whose mapped distribution shows that they occur in less than 16 of the 10×10 km squares into which Ordnance Survey maps are divided in Great Britain. Studies by Susan Bye (now Mrs Joy) showed that if this criterion were adopted the number of sites would be excessive and so a more complicated but more rigorous system had to be proposed. First we agreed that all the localities of these Red Data Book species which were known to be under threat and therefore had been listed in Schedule 2 of the Conservation of Wild Creatures and Wild Plants Act 1975 should be notified. In addition we devised a system by which outstanding assemblages of plants should also be notified. An outstanding assemblage was defined as one which had an index of 200 or more. Plants occurring in 31–60 map squares of 10×10 km^2 scored 40, those in 16–30 squares scored 50, and those in 15 or less 100. This ensured that all sites containing two Red Data Book species would be notified, and that those with one Red Data Book species accompanied by other rare species would also be included.

As we have seen, most SSSI were to be selected as the best examples of different habitats in counties. One of the first questions we had to ask was, whether it was feasible to carry out primary surveys of all habitats within a county from which to select the best examples. Pilot studies by Mrs Helen Hepburn and Miss Janet Forbes showed that not less than ten botanists would be needed to cover the county of Kent in one field season. Grasslands provided the most difficult problem, since no distinction between ancient grasslands and grass leys is made on maps. Janet Forbes tried to relate aerial photographs with what she saw on the ground. Grasslands which were being invaded by scrub could be identified easily and many of these were in fact old grasslands. On the other hand, the subtle distinctions between many old grasslands and leys of varying ages could not be determined from aerial photographs. This meant that all grasslands had to be visited on the ground. The study in Kent suggested that three-fifths of the time spent by a survey team of ten botanists in that county would have to be devoted to the survey of grasslands.

The Kent study showed that practically all the areas already notified as SSSI would have been confirmed by the criteria proposed by the working group. However, it also showed that numerous other sites would have to be notified as well if the criteria were adopted. Most of the new sites would be deciduous woodland, grassland and open-water habitats.

Meanwhile similar studies were made in Bedfordshire and Argyll, and these also showed that the vast majority of existing sites were confirmed by the new proposals, but that further sites would have to be added.

As the notification of SSSI developed financial consequences there was increasing concern about where their boundaries were drawn. They had to include the area of special interest, but how much else? A site for a rare orchid clearly had to consist of more than the few square feet in which the orchids grew – the whole field or wood providing the habitat had to be notified. If developments on land immediately surrounding the habitat could destroy its scientific interest should not the site include some bufferland so that the Nature Conservancy Council would be consulted about such developments? Over the years, Regional Officers had used their common sense about this matter. If the scientific interest of a valley bog was restricted to its upstream end, the whole bog was notified because a drainage scheme downstream of the place of special interest would destroy the latter as effectively as a scheme on the site of the special interest itself. However, many expressed concern that the boundaries of SSSI might have been drawn too liberally in the past and so a special study of existing bufferland was undertaken by Susan Bye. She studied the problem in Wales, which contained numerous large SSSI with mires and open water, and so was likely to have more bufferland than many parts of Britain. She tested the guideline on bufferland for regional staff which we had proposed. This guideline stated that SSSI should include that land 'in which changes in land use are likely to destroy or severely damage the scientific value of the site which it surrounds'. She showed that, far from there being too much bufferland in Wales, there was too little. In the sample of seventy-three SSSI only twenty-three contained bufferland, but thirty-nine should have contained it. In many of these, the SSSI had already sustained damage because it had not contained bufferland. Susan Bye further concluded that if the guideline adopted by the working group were adopted, only 2% of the total area of SSSI in Wales would consist of bufferland. This could not be considered excessive.

SSSI consisting of whole rivers or parts of them posed a special problem since pollution upstream can always destroy the interest downstream, and thus ideally the whole catchment area should be included as bufferland in a river SSSI. This would be impracticable in most cases. Thus, in the case of river SSSI, separate measures have to be taken with the Water Authorities and others to ensure that the risk of pollution is reduced as far as possible without resource to notifying the whole catchment as a SSSI.

The guideline proposed by the working group took note of all these studies and they were submitted to the Council of the Nature Conservancy Council in 1979 for approval. Since that date they have formed the basis for revising old sites and selecting new ones.

The introduction to the guidelines made it clear that they must be kept under continuous review, not only because of new scientific knowledge but because of changes in the countryside as a whole. My final task for the Nature Conservancy Council was to complete a revision of the guidelines in 1983. It will not be the last.

Administrative duties are essential, but they tend to seem dull and uninteresting unless one is personally involved in them. Then they become exciting, full of the conflict between ideas and between people and full of drama. The selection of SSSI is no exception. It is a practical, administrative matter yet it lies at the heart of conservation in Britain, linking science and politics, man and nature.

The choice of SSSI depends essentially on:
(1) the determination of conservation objectives – what are we aiming to do?
(2) scientific knowledge about habitats and species, and
(3) the evaluation of the respective conservation roles of sites and unnotified habitats in the wider countryside, within the nation's economic and political framework; this, in turn, depends on the constraints provided by the world in general and the EEC in particular.

There is much room for argument about the objectives of the SSSI system because people, not least professional conservationists, differ on the detailed objectives of conservation. There is room for argument about scientific matters because information about the distribution and vulnerability of habitats and species is still very incomplete. Therefore no one can afford to be dogmatic about the selection of SSSI. Yet the notification of SSSI is a statutory duty of the Government's official conservation organisation, the Nature Conservancy Council. Therefore decisions have to be made using the best information available. In the next chapter I shall try to put the conservation of special sites, both National Nature Reserves and SSSI into their wider context.

8

*

Farms and farmers

Networks of nature reserves and SSSI are essential if we are to conserve the flora and fauna of Britain because they provide relatively secure bases for all our plants and animals. When opportunities for colonisation arise, many species of wildlife can spread out into the surrounding countryside from these bases.

Rare species usually depend on special conditions which can only be maintained in protected sites, and so their survival depends increasingly on nature reserves and SSSI. Rare species need special measures because they are threatened, not because they are more valuable than common ones. It cannot be said too often that it is as much the conservationist's job to keep common species common as it is to ensure the survival of rare species. Even if the gene pool could be conserved in nature reserves and SSSI alone, a state of affairs which denied people contact with wildlife in ordinary life and ordinary places would be grossly deficient. For most people, an abundance of the relatively common and conspicuous plants and animals is more important than the conservation of obscure rarities.

The conservation of wildlife habitats in the wider countryside is thus important both for scientific and social reasons. First, it contributes to the conservation of the flora and fauna of the country as a whole. It does this by providing extra habitats over and above those found on nature reserves and SSSI, and by providing 'stepping stones' and 'corridors' which link scattered populations and hence help to ensure their survival. Secondly it keeps common species common, so that those who live in or visit the countryside can see them and enjoy them.

To say that nature reserves and SSSI are primarily for conserving rare species while habitats in the wider countryside are for the common species is a gross simplification. It does, however, have some truth and emphasises a very significant point; it underlines the importance of the conservation role of those who manage the wider countryside. Specialist conservationists working on protected sites and farmers, foresters and landowners working in the wider countryside complement each other fundamentally. Both groups of people are necessary for conservation and it is vital that both should recognise this, since they need to cooperate in the common cause of maintaining our living resources on behalf of society as a whole.

Conservationists have a dual role concerning land. On the one hand they have to acquire nature reserves and manage them so that they continue to produce their 'crops' of special wildlife. On the other they have to advise those who earn their living by farming and forestry about how to support wildlife as a side line to their ordinary business.

The staff of the Nature Conservancy Council and members of the Nature Conservation Trusts have frequently given advice to individual farmers about conservation in the wider countryside, but the numbers of advisers are so small and the need to conserve nature reserves and SSSI is so great that the direct impact of conservationists on the conservation of habitats in the wider countryside has been slight. The statutory duties of the Nature Conservancy Council concerning sites are so onerous that they have forced the Nature Conservancy Council to be preoccupied with them. Despite the crucial role of farmers and foresters in conservation, much too little time has been devoted by the conservation organisations to advising them. This has been unavoidable in the circumstances but it is a great pity, since no group of people can do more for conservation than farmers, the people who manage most of the countryside. Conservationists and farmers have been kept apart when they should have been working together. The gap is being narrowed by the development of the cooperative movement known as the Farming and Wildlife Advisory Group (FWAG).

The development of FWAG is particularly relevant to the

themes of this book since it illustrates how ideas about conservation have evolved among the people who can do most about it. I have had excellent opportunities to observe the growth of FWAG as I was a founder member and was its chairman at a critical stage in its development. How did it arise in the first place?

In earlier chapters I have described how the public was first made aware of habitat loss. Inevitably this had an effect on the public's image of the farmer. In the past he was considered to be the guardian of the countryside, then increasingly he appeared in the guise of its despoiler. The facts about habitat loss (and, as we shall see, about the effects of pesticides) had to be exposed, but the antagonism which they generated was clearly bad for conservation because the farmers' cooperation was crucial for conservation. Therefore in the late 1960s a small group of us got together and planned an exercise which would bring farmers and conservationists together in order to discuss practical conservation measures on a farm. The underlying assumptions were important. We did not deny that the requirements of modern farming often ran counter to conservation, nevertheless those requirements were real. We also assumed that farmers in the main did want to conserve wildlife and maintain a beautiful landscape, but the extent to which they could be effective was very much influenced by their financial situations.

The exercise took place on a farm owned by the *Farmer's Weekly* near Tring in Hertfordshire. However, the participants of the exercise stayed nearby at Silsoe College in Bedfordshire, and so the conference afterwards became known as the Silsoe Conference.

Before the exercise was run, biological surveys were made of the farm so that we started off knowing what there was to conserve and where it was. As with most farms, the wildlife on it was unevenly distributed. The best places included a small remnant of chalk grassland and the edge of a canal lined by large bushes (Fig. 29). As on most farms, the hedges differed in age and hence in composition and value. At least one pond must have been good for wildlife in the past, but had lost much of its value through neglect.

On the day of the exercise (12 July 1969) we divided up into six parties each with different agricultural and conservation tasks.

We were all made aware that if the owner had inherited the farm he could do a good deal for wildlife; whereas, if he had just bought it, and thus had to intensify his farming to cover a large mortgage, he simply could not afford to do any conservation which cost much money. We then walked the farm and decided on what best could be done within the constraints of our particular remits. In the evening each syndicate leader outlined his proposals. My job was to assess the biological effects of the proposals. There was so much information available that I had to burn a lot of midnight oil in order to have my conclusions ready for the following day.

Everyone who attended the Silsoe Conference learnt a great deal – the farmers about conservation, the conservationists about farming. The exercise showed that the problems of conserving wildlife on a modern farm were real and were difficult to solve, but we all left feeling that they were not insurmountable

Fig. 29. Good habitat on the site of the 'Silsoe exercise' 1969. As on all farms, some parts were much more valuable for wildlife than others. The Silsoe exercise led to the formation of the Farming and Wildlife Advisory Group (FWAG).

so long as farmers and conservationists worked in partnership. The Silsoe exercise showed that they could.

At the end of a successful conference there is always a feeling of sadness. New friendships are made and new possibilities of cooperation appear and seem likely to succeed, and then suddenly everyone disperses and the glimpse of progress abruptly disappears. The Silsoe experiment seemed too valuable to lose. We were determined that it should continue in some form, so we founded the Farming and Wildlife Advisory Group.

At first, FWAG did little more than provide a national talking shop for members of the main organisations involved with the management of land – the Agricultural Development and Advisory Service of MAFF, the British Trust for Ornithology, the Country Landowners' Association, the Countryside Commissions, the Forestry Commission, the National Farmers' Union, the Nature Conservancy Council, the Royal Institute of Chartered Surveyors, the Royal Society for Nature Conservation and the Royal Society for the Protection of Birds. We employed Mr Jim Hall, who had farmed in Cambridgeshire, as our adviser. Initially his job was to organise national exercises of the Silsoe type in other parts of the country, but soon we felt that the way ahead lay in developing FWAGs at the county level. These were designed to do for the county what national FWAG did for the country as a whole; they were to build up a partnership between farming and conservation interests. Farmers made up an important part of their membership. Local FWAGs were encouraged to take the next and vital step – to give practical advice on conservation when they were asked for it. Jim Hall did a tremendous job in getting the local FWAGs started, and his successor, Mr Eric Carter has continued it most effectively. From being a talking shop FWAG was becoming a cooperative agency for promoting such things as coppicing woodland, hedgerow and grassland management and digging ponds.

A very successful experiment in Gloucestershire and Somerset, which was financed by the local Naturalist Trusts, the Countryside Commission and the Nature Conservancy Council, showed that if each local county FWAG could employ a paid full-time adviser many more farmers sought advice from the county FWAG and far more conservation was achieved. As a result of this experiment, the Farming and Wildlife Trust has been set up as a charitable trust in order to finance county advisers throughout the country. By early in 1985 it had raised £1.3 million. There are sixty-two county FWAGs in England, Scotland and Wales and over 30 have full time advisers. It is too early to assess the full effectiveness of FWAG, but it shows great promise and already many old woods, new plantations of native trees, hedges and ponds throughout Britain owe their existence or effective management to FWAG.

I think that all of us who have worked with FWAG would agree that we have learnt a great deal in the process. We can understand why many farmers appear to be destructive phil- istines to many conservationists, and why many conser- vationists appear to be impractical busybodies to many farmers. The differences in perception are very real, as is to be expected when two lots of people look at the same thing from quite different viewpoints. At the risk of caricature I shall try to describe the two viewpoints.

The farmer today is essentially a businessman trying to make as large a profit as he can out of his capital investment in land and machinery. To that end he tries to produce as much as possible from his land, with all the aids provided by agricultural science which he can afford. He looks at wildlife primarily from the point of view of 'will it affect yields?' In other words, he is mainly concerned with those wild species which either compete with his crops for space or those which reduce yields by eating them. Thus the wildlife which most concern farmers are the baddies – the weeds, fungal diseases, insect pests, pigeons and rats. If the farmer is a shooting man, partridges, pheasants and, to a lesser extent, hares and wildfowl, are goodies; the vast majority of species (which mainly live on the uncropped parts of the farm) are largely ignored unless the farmer is a naturalist. He will notice increases and decreases of pests and game but be unaware of what happens to that vast majority of 'neutral' species. Highly productive crops of cereals or grass are by their nature uniform and hence tidy. Tidiness becomes a symbol of good farming, and indeed there is great beauty in great rolling expanses of well- grown corn. In the past, high productivity was achieved by carefully relating cropping to minor differences in soil and drainage so the farmer had to know his land in detail and he often got clues about it from wild plants. Today small differences can be eradicated by deep draining and chemical treatment and so there is less need for the farmer to look at the small print of his land. This further distances him from its ecology. He and his farm workers are also distanced physically from it since they travel about it in machines, and most operations are carried out by machines rather than by hand. There is much less physical contact with nature; farmers are frequently far less knowledgeable about wildlife than interested visitors from towns, or commuters who work in towns but live in villages. Even the most mechanised farm still supports a lot of wildlife, although it is largely made up of a large number of individuals of a few very robust species such as house sparrows, cabbage white butterflies and couch grass. The farmer who is not interested in wildlife still sees a lot of it from his tractor – too much of it he may think. He finds it difficult to see what all this conservation nonsense is about.

Those who do not earn their living by farming, but are interested in wildlife, look at farms in quite a different way. To them, the farm is a bit of countryside made up of good and not-so-good wildlife habitats. Naturalists scarcely look at the crops because they are no longer interesting from a naturalist's point of view. They find what they are looking for in the untidy parts of the farm – the woods, the hedges, the ponds and the few remaining fields which have not been treated with herbicides or fertilisers. They yearn for the past when the croplands themselves, especially the grasslands, supported numerous species of wildlife. While the farmer thinks of pesticides and fertilisers as means of increasing yields, the visiting naturalist views them as agents of destruction of the things he has come to see.

These two very different perceptions of the countryside are reinforced by strong social factors. Farmers tend to be relatively well off and they live in very pleasant places: as a result many of the visitors from the towns regard them with envy and think of farmers as 'them', not 'us'. When one considers the differences in perception and the differences in day-to-day experience, it is not at all surprising that cooperation between farmers and urban conservationists is not an easy matter.

The important fact, and the key to the solution of the problem is that the stereotypes described above do not cover everybody. There are many farmers who care a great deal about conservation and have been managing their land to support wildlife for many years without any promptings from conservationists. There is a growing number of conservationists who know enough about farming to realise perfectly well why farmers act as they do and are sympathetic towards them. We can build on the experience of such people, and FWAG has.

Nobody does anything effectively unless he or she is strongly motivated to do it – this applies to farmers no less than the rest of us. If a farmer couldn't care less about wildlife no amount of advice, cajoling or legislation will make him conserve wildlife effectively; once his imagination has been caught and he wants to conserve what is on his land, he will ask the right questions. He will ask what the species requires, he will ask how its requirements can be provided and the cost of so doing. Thus the conservationists in FWAG have a dual role. First they must help interest their farming colleagues, and second they must give sound practical advice on habitat management and be able to cost it. From then on, conservation becomes a partnership.

The degree to which this partnership can flourish will depend to a large extent on the fiscal and economic framework within which it has to operate. At the moment the Common Agricultural Policy (CAP) could hardly provide a less favourable environment for trying to combine conservation with farming, yet FWAG has shown that, even under the adverse conditions of today, real partnership leading to effective conservation can be achieved. This is an extraordinarily hopeful conclusion.

FWAG is merely an umbrella organisation, under which those concerned with the management of land can cooperate in order to promote conservation on farms. Those of its members who are not farmers come from different organisations, whose individual views differ considerably. This prevents FWAG from making pronouncements about important but controversial topics such as the CAP, straw-burning or aerial spraying. It is no accident that, while its members come from different organisations, they do not represent them. How is it that this amorphous, voiceless body has been so successful in catching the imagination of the farming community and in getting practical conservation work done? I suspect that its virtue has stemmed from a necessity, that FWAG owes its success to its neutrality.

This allows it to be a conservation body with which farmers and those that support them can identify.

It is worth considering the development of FWAG in the context of time because it has much to teach us. FWAG has been most fortunate in the timing of its expansion. The formation of the Farming and Wildlife Trust in 1984, which enabled it to finance its network of county advisers, coincided with the very moment when the general public came to recognise the connection between the loss of wildlife and the butter and cereal mountains and, as a result, became less kindly disposed towards the farming community. In these circumstances, those farmers who had always been good conservationists looked with favour on a conservation body which recognised their efforts. At the same time other farmers, who had not taken conservation seriously until then, became unhappy about the unnecessary damage to wildlife and landscape and about their image in the eyes of the public. They too wanted to demonstrate that they did care and were doing something about it. They wanted to show that conservation was indeed their thing and not just something foisted on them by the non-farming public. FWAG has not been afraid or ashamed to cash in on these special circumstances of the 1980s. One can speculate that it would have been much less successful in getting financial and moral support if it had launched its appeals before the general public had become dissatisfied with the working of the CAP.

The aim of FWAG is to get farmers to make the best conservation use of their land within the constraints of efficient farming. This is achieved by giving sound advice on the management of existing wildlife habitats on the farm and on making new ones. On conservation grounds, retaining and managing the existing habitats is usually a much more important activity than making new ones, yet FWAG has discovered a much greater willingness by farmers to plant trees and dig ponds than to coppice existing woods and resuscitate old ponds. There have been too many instances when a farmer has cut down a wood and taken out an excessive number of hedges and has salved his conscience by planting a few field corners with trees. Conservationists have rightly complained when they have seen this happen, but they should not condemn FWAG as a result. FWAG has made its priorities very clear. It recognises that the maintenance of existing woods is more important than planting new ones. However, FWAG is involved in a long-term campaign of education, and experience has shown that the first and critical contact between itself and the farmer is usually provided by tree planting. FWAG advisers must respond to that request and build positively on that foundation.

The loss of trees through Dutch Elm Disease has made thousands of farmers aware of the value of something which they had hitherto taken for granted. Once the decision to replace the elms had been taken, questions had to be asked and answered. Which species should be planted? Where should they be planted? How much care would they need after planting? These and other questions lead to an interest in the trees and often to the wildlife that makes use of them. This, in turn, leads to an interest in the wildlife on the farm as a whole. At this stage, the farmer realises why the old established habitats, which he had hardly thought about before, are so important. Everything begins to make sense in an evolutionary and historical framework. Conservation becomes meaningful as well as a delight.

One of the practical problems of conservation on the farm is that it takes time to get results. Therefore any conservation activity which quickly produces dividends is very welcome. As we have seen, one of the delights of digging ponds is that they quickly become colonised: one gets results straight away. So digging ponds is a particularly good way of stimulating an interest in wildlife and conservation.

I believe that all conservation activities on the farm have a significant symbolic element in relation to time. It is most obviously apparent in tree planting. Trees live longer than men and no man can plant an acorn and see the oak tree grow to maturity. Therefore to plant a tree is an act of faith. Once this is recognised by the traveller, a landscape full of young trees becomes an inspiring symbol of hope which is the necessary base for conservation.

9

*

The habitat crisis and the involvement
of government

Conservation must always be a combined operation between
the private sector and the public. In this chapter I shall
describe some attempts which were made to improve the system
so that government could make it easier for the Nature Conservancy Council, other conservation bodies and private citizens to
conserve habitats.

I have described the three ways in which conservation of
habitats is undertaken in Britain. The establishment of National
Nature Reserves is mainly a matter for government. Public
money is spent on selecting the reserves, on buying or leasing
the land and on managing them.

SSSI are also selected by the Nature Conservancy Council, the
government's conservation agency, but they are mainly owned
and managed by private individuals or organisations.

Most wildlife lives in habitats over which government has
little or no direct control, although government policy on agriculture and forestry has profound indirect effects on the way these
habitats are regarded and managed by individual farmers and
landowners.

The practice of conservation during the 25 years following the
formation of the Nature Conservancy showed us the strengths
and weaknesses of the system under which we worked. National
Nature Reserves could be bought, leased or managed under
Nature Reserve Agreements. This flexibility enabled the Nature
Conservancy to control much more land than would otherwise
have been the case. On the other hand, the amount of money

available for reserves in any one year was inadequate at a time when many potential nature reserves were coming under threat.

As we have seen, the SSSI procedure gave useful protection of sites from developments which required planning permission, but initially lacked any mechanism by which other kinds of development could be prevented or modified. As modern methods of farming and forestry became less suitable for SSSI, the cost of managing SSSI in ways which retained their conservation value became greater, yet landowners received very little financial support for managing them properly.

Lip service was given to the need to conserve wildlife on all land. Under the provisions of the Countryside Act 1968, ministers, government departments and public bodies were always 'to have regard to the desirability of conserving the natural beauty and amenity of the countryside', and it was made clear that this included 'flora, fauna and geological and physiographical features'. Nevertheless, the whole system of grant aid went in the opposite direction: it encouraged farmers to increase productivity regardless of the consequences to the environment. Many hedges were removed, much wetland drained, many marginal grasslands improved and many woods with native trees converted into conifer plantations, simply because public money was available to do these things. Of course, increasing productivity was desirable for economic and social reasons, but the system of grant-aid support was so crude that sensible compromises between food and timber production and conservation were extremely difficult to implement.

By 1975 it was evident that a crisis point had been reached. Surveys of habitats and of individual species had shown unprecedented declines. The lack of money for acquiring reserves, and the declared aims of government to maximise food production had shown that the crisis was not perceived outside conservation circles. The time had clearly come for the Nature Conservancy Council to warn the public in general, and the government in particular, about the consequences of inaction, and to emphasise the need to change the rules so that government could give more support for conservation. Both the Nature Conservancy Council and the voluntary conservation bodies and individual farmers needed help so that they could conserve wildlife habitats under the new conditions. The Nature Conservancy Council realised that the conservation problems both of sites and of the wider countryside could not be solved by tinkering with existing arrangements: they called for a radical new approach to conservation by the nation.

I had recently left Monks Wood and had been appointed to the new post of Chief Advisory Officer to the Nature Conservancy Council. My first main task was to study the impact of agriculture on nature conservation – since that lay at the root of our problems – and to make recommendations. This was an exciting assignment and I was able to make use of my previous experience by drawing together the various threads of my career. My immediate task was to prepare a discussion paper in order to involve all the government departments and non-governmental organisations which were concerned with land management in Great Britain.

We could not expect others to make changes in their procedures and activities unless they accepted that there really was an urgent problem to be solved. My experiences with habitat conservation had shown me that most people did not appreciate either the scale of habitat destruction or the urgency to react to it. Therefore in the discussion paper I had to ask and answer such questions as why does wildlife matter? What particular effects does the intensification of agriculture have on wildlife? Are we losing significant amounts of wildlife habitat? Are there not plenty of nature reserves: if so, why worry about farmland? Why cannot the conservation of wildlife be confined to marginal land in the uplands? Do gains in new habitats compensate for losses in old ones? Are existing procedures effective? The paper showed that there was good evidence that wildlife habitats had declined significantly and that this was confirmed by data on the distribution of British plants, which had been collected over the years by the Biological Records Centre at Monks Wood. For example, the number of 10 km squares containing populations of Britain's rarer plants had been reduced by 30% between 1930 and 1960. The loss was greatest in the very rare species. In 1900, forty-four species were so rare that they only occurred in one or two 10 km squares; by 1970 the number of such species had risen to ninety-seven. It showed that there were no easy solutions. Nature reserves alone could not support our flora and fauna; new habitats nowhere near compensated for the loss of old ones;

conservation could not be relegated to marginal land in the uplands because many species were confined by their genetic requirements to the lowlands. All this set the scene for recommendations. The main ones were:

(1) special emphasis should be put on establishing nature reserves in the lowlands, where the threat to habitat was greatest;

(2) formal consultation on change of land use in SSSI should be extended to cover agricultural and forestry operations;

(3) owners should be given greater financial incentives to manage SSSI so that they retained their scientific interest;

(4) grant aid for conservation should be extended to wildlife habitats in the wider countryside;

(5) the Ministry of Agriculture, Fisheries and Food and the Scottish Colleges of Agriculture should take a positive role in promoting conservation through their advisory services;

(6) government departments and agencies should set an example to other landowners as regards conservation management;

(7) the most important recommendation was that

a forward-looking rural strategy for our national resources should be formulated, which recognises among other things that wildlife is a vital part of the real capital wealth and heritage of the nation. Such a policy would provide the framework within which government departments could determine their priorities and could seek to optimise all situations in the national interest.

The discussion paper was circulated to thirty-nine government departments and non-governmental organisations for comment. There followed much debate within the Council of the Nature Conservancy Council about their final pronouncements. Knowledge of conservation is not a necessary qualification for members of the Nature Conservancy's Council. Some members have considerable knowledge but others are appointed for other reasons. The latter were typical of many people whom we were trying to influence. It took a great deal of time and argument to convince them that the plight of wildlife and their habitats was as bad as we said it was, and so could only be remedied by the measures we proposed. Eventually agreement was achieved and in 1977 the Nature Conservancy Council published its recommendations in *Nature Conservation and Agriculture*. This booklet included summaries of the comments received in the 1976 discussion paper. The Nature Conservancy Council's stance, which received a great deal of support from the organisations consulted, was summarised on the cover of the report:

The Nature Conservancy Council is convinced that we must now be prepared to plan, and indeed to pay, to maintain our heritage of wild plants and animals. Its recommendations include a call for a rural land use strategy that takes account of nature conservation as well as the need for increased food production and other national requirements. This will necessitate improvements to current regulations, advisory services and grant aid, to assist the individual to manage his land in the best interests of the nation without financial loss.

With one important exception, the recommendations were wholly or partly implemented less than 4 years later, when the Wildlife and Countryside Act became law. Owners and occupiers do now have to consult with the Nature Conservancy Council about farming and forestry operations on SSSI, and more money is available for the management of SSSI for conservation; MAFF and the Scottish Colleges of Agriculture are exploring ways of coordinating conservation advice with the Nature Conservancy Council and FWAG advisers. Some government agencies, notably the Foresty Commission and the Ministry of Defence, are giving considerably more thought to conservation on their extensive holdings. More cash has been made available to the Nature Conservancy Council and the number of lowland and upland National Nature Reserves has been increased.

Sadly and significantly, the principal recommendation made in the Nature Conservancy Council's document *Nature Conservation and Agriculture* has not been implemented: there is still no national strategy for land use in Britain. As a result, each contentious case is taken on its local 'merits', without reference to a national plan because no such plan exists. Even today good farmland is sometimes used for industrial development when derelict land nearby is left derelict. Important wildlife sites are still being destroyed in order to produce unwanted surpluses of dairy and cereal products. Public money is wasted on public inquiries which have to be held because there are no simple guidelines on land-use priorities.

Careful consideration of how we use our limited resource of

land is as necessary for landscape conservation and hence for recreation and the tourist industry as it is for nature conservation. Sir Derek Barber, the chairman of the Countryside Commission, has recently made a strong plea for strategic thinking about land, but to date he seems to have had no better response than the Nature Conservancy Council got in 1977. Why do governments which applaud forward planning in industry jib at strategic planning for land? A land-use strategy neither implies state ownership nor excessive control of enterprises. Yet I believe the concept of a rural land-use strategy has been opposed principally on the ground that it would be the thin end of a land-nationalisation wedge. Also, some people are so impressed by the multiplicity of special circumstances that they cannot imagine how rather generalised guidelines could really help. Of course local constraints should be considered in every case, but not given more weight than the national good. It must be right to conserve our best land for agriculture and our best habitats for wildlife. A land-use strategy does no more than make those presumptions. They are crucial if we take our living resources seriously. I shall return to this theme when considering what action should be taken in the future.

Habitats are most profoundly affected by changes in land use, but they can suffer deterioration from more insidious causes. The next part of this book describes new hazards which affected special sites and the wider countryside alike, and which impinged on people and government to an unprecedented extent.

PART III

---- ★ ----

The past: experience from controlling disease and pollution

10

*

The consequences of a disease

Why is that rabbit sitting exposed on the path? It seems unnaturally tame – is there something wrong? As we get nearer, it does not run away but turns towards us and blunders into a clod of earth. It is blind, and now that we can see its dreadfully swollen head we can understand why; it has got myxomatosis. Should we kill it or leave it to die? Other questions must be asked.

Conservation of habitat is the *sine qua non* of nature conservation. Yet wildlife habitats, no less than crops, can be seriously affected by disease and pollution operating indiscriminately over the earth's surface. Therefore conservationists have to promote the control of disease and pollution as well as the conservation of habitats. Today this is all too obvious: the devastation of much of the English landscape by Dutch Elm Disease is there for everyone to see. A failure of quarantine measures in the timber trade not only destroyed the landscape, but must have caused catastrophic declines in the populations of species (such as that of the White Letter Hairstreak Butterfly) which are largely dependent on elms for their food. Birds and bats dependent on holes in trees lost most of their breeding sites in those districts where hedgerow trees were predominantly elms. At the same time, large areas of western Europe (including Britain) are being affected by industrial pollution in the form of acid rain. Today more and more people realise that we must make much greater efforts to prevent pathogens from entering the country, and we must take much tougher measures to control pollution.

When the Nature Conservancy was set up in 1949, damage by local hot spots of pollution were regretted – an oil spill here, a discharge of poisonous factory effluent there, but no one realised how important environmental contamination on the large scale would become. The arrival of myxomatosis in 1953 introduced us to a problem which affected the whole country. This was a new experience and we learnt from it.

By curious coincidence, the arrival of the myxoma virus in Britain coincided with the first recognition that pesticides could affect wildlife significantly. Thus the environmental problems of both biological and chemical control of pest species arose simultaneously. In so far as one can put a date to such things, 1954 was the beginning of the present conservation era, in which conservation was seen to be a matter both of site protection and of the general protection of the environment from disease and pollution.

The European Rabbit (*Oryctolagus cuniculus*) was originally a native of southwestern Europe. It was valuable for food and fur and was introduced into England by the Normans. Subsequently it was introduced into numerous other parts of the world, largely as a sporting quarry. When the wool trade dominated British farming and huge areas were under grass, the rabbit increased greatly in Britain. It thrived, particularly in times of agricultural depression. Before 1954 it was so numerous that it was a major ecological factor: it determined which species occurred or predominated in many grasslands and prevented natural regeneration of woodland in many districts. Yet both farmers and conservationists tended to underestimate the ecological importance of the rabbit: it was taken for granted.

The myxoma virus, a pox virus, was discovered by G. Sanarelli in 1897. It was endemic in the Cottontail Rabbits (*Sylvilagus*, see Fig. 35) of South America. The virus was in equilibrium with its host: an affected Cottontail only developed local lesions and was never killed by the disease. In 1927 Professor Aragao of Brazil suggested that myxoma might be used for the biological control of the European Rabbit in areas where it was a devastating pest. Subsequently, in 1933, Sir Charles Martin carried out feasibility trials at Cambridge on behalf of CSIRO in Australia. These laboratory trials showed that the virus was extremely effective in killing European Rabbits. However, field trials by Mr Ronald

Lockley on the Pembrokeshire island of Skokholm and others in Denmark were not successful in controlling rabbits. In 1950 the combined effects of a drought and rabbit grazing turned large areas of Australia into desert. This stimulated the introduction of the myxoma virus into the Murray River area. It spread quickly by means of a mosquito vector. The results were spectacular – the rabbit population was greatly reduced to the great advantage of Australian farmers – £50m gained was the sum mentioned at the time. Also, numerous wild species benefited greatly since the vegetation recovered and competition from rabbits virtually disappeared.

In June 1952 Dr Armand Delille, a French doctor, introduced the myxoma virus onto his estate near Paris. By 1953 the disease had spread to most of France, and into Luxembourg, Belgium, the Netherlands, Germany and northern Spain. Dr Armand Delille's action caused a tremendous uproar and litigation. He was supported by numerous farmers but attacked by the chasseurs who had lost one of their main quarries.

In October 1953 myxomatosis was reported at Edenbridge in Kent. It appeared later in East Sussex, Essex and East Suffolk. By the end of 1954, it was well established throughout England and Wales and had reached Scotland and Ireland. Some believed that myxomatosis had reached Britain by means of migrating birds, others that it was deliberately introduced. I believe the latter to be the case although no one has yet admitted to introducing the disease. Once here, the disease was widely disseminated by farmers wishing to control rabbits, so it was difficult to distinguish the effects of natural spread from artificial introductions. Mr Harry Thompson showed that natural spread was slow. In 1954 experiments by Ronald Lockley showed that the vector of the disease in Britain was the flea *Spilopsyllus cuniculi*. That explained why his experimental introductions to his farm on Skokholm in 1936 and 1937 did not work; the Skokholm rabbits do not have fleas!

As early as 1954, Ronald Lockley suspected that the virus was becoming less virulent and therefore more rabbits were surviving infection. This process was aided and abetted by natural selection because the longer a diseased rabbit survives the greater is the chance of it supporting fleas containing that form of the virus. Attenuation of the virus has continued in Britain and elsewhere. Much of the control work in Australia consisted of spreading the more virulent strains of virus artificially. More recently there are signs that the rabbit itself is becoming resistant to the disease. Nevertheless, despite these trends, the rabbit is far less abundant in Britain today than it was in 1953 when myxomatosis was introduced.

When it became clear that the initial outbreaks of myxomatosis could not be contained it seemed likely that the rabbit, one of Britain's commonest mammals, would become very much rarer if not extinct. Enough was known about its effects as a grazing animal and its importance as prey to many predators for the Nature Conservancy to be deeply concerned about what might happen. The Nature Conservancy had to have the latest information about the disease and its spread. It had to record the effects of this unique event as far as possible, and it had to be in a position to influence government policy.

Liaison with the Ministry of Agriculture was quickly established through my appointment as an observer at meetings of the Advisory Committee on Myxomatosis and as a member of its scientific sub-committee. These arrangements provided a formal link between the Ministry of Agriculture and the Nature Conservancy.

The most fundamental ecological changes resulting from a catastrophic decline in the rabbit population were likely to be botanical. My colleague, the late Dr A. S. Thomas was given the task of recording the changes on sample nature reserves and other sites. He did this both by recording plants on transects and by taking photographs from the same points in 1954 and 1955. 7265 m (23 837 feet) of transects on ten sites were recorded. In the space of one year, Dr Thomas reported 'a greater variety of species, a greater height of turf and a better cover of the ground'. The disappearance of rabbits 'had a spectacular effect on the show on the downs this year both of common plants like the cowslip, ragwort and rockrose, and of rarer plants such as the Pasque flower and some of the orchids'. However, Dr Thomas warned that 'where the turf is not controlled by increased stocking coarse grasses and woody plants may crowd out some of the more interesting small herbs'. Subsequent work confirmed these fears.

Dr Thomas's transects at Horn Heath in the Suffolk Breck, at

Bromhill Burrows and Skomer Island in Pembrokeshire were done to supplement the studies on rabbit populations which were being carried out by Harry Thompson of the Ministry of Agriculture and Ronald Lockley.

I felt strongly that studies on the indirect effects of myxomatosis, should not be confined to plants, and so I explored the possibilities for studying the indirect effects of the disease on a variety of species of animals. I listed species which appeared to depend on rabbits for food or habitat and hence could be expected to decline as the result of a catastrophic decline in the rabbit population. I also listed species which appeared to compete with rabbits for grazing and which might increase as more food became available to them. I discarded most species for study on the grounds that they were too rare or too difficult to observe. My short list of species which must be at least partly dependent on the rabbit consisted of Fox, Badger, Stoat, Weazel and Buzzard (all of which preyed on the rabbit) and the Minotaur Beetle (*Ceratophyus typhoeus*) (Fig. 30) whose larvae feed on rabbit dung pellets, the Wheatear which nests in rabbit holes and the Stone Curlew which was at least partly dependent on bare ground produced by excessive rabbit grazing. My list of possible competitors consisted of Brown Hare, Mountain Hare, Red Deer, Roe Deer, and Short tailed Field Vole. I hoped to get

Fig. 30. Minotaur beetle (*Ceratophyus typhoeus*). This beetle lays its eggs in burrows in the sand. The male provides the larvae with rabbit pellets which it finds on the surface near the burrows. The species survived the dearth of rabbit pellets following the introduction of myxomatosis (enlarged).

some of the information I required by studying game books and records of vermin kills. The hunts might be able to help with Foxes, the Bureau of Animal Population with voles; but work on birds would depend on special studies, and hence on collaborative work with ornithologists. It was soon obvious that even superficial studies on the species in my short list made up a programme that was far too ambitious and so I restricted my proposals to one predator, the Buzzard (Fig. 31), and one competitor, the Brown Hare. These species were particularly convenient for my purposes since the Buzzard was particularly common in the region in which I worked, and the Hare was unevenly distributed there and so changes in its distribution might be more apparent in the South West than in districts where it was already universally common.

When I sought permission to study the inherent effects of myxomatosis on the Buzzard I encountered an unforeseen problem. The resources of the Nature Conservancy were very limited and, since research on birds was being done by the Edward Grey Institute and by the British Trust for Ornithology (BTO), Nature Conservancy officers were not encouraged to work on birds.

Fig. 31. Buzzard (*Buteo buteo*). Rabbits were formerly the principal food of this species. Myxomatosis had a marked indirect effect on its breeding success until it became adjusted to feeding on alternative prey.

Therefore I had to convince Max Nicholson, the Director of the Nature Conservancy, that we should make an exception in the case of the Buzzard. After some discussion he agreed that I could organise a BTO survey of the Buzzard to determine the effects of myxomatosis on its feeding habits, reproduction and distribution.

With the help of Dr Bruce Campbell, the business editor of the newly established journal *Bird Study*, the survey was advertised and questionnaires were sent off to all who had offered help in time for the 1954 breeding season. Those taking part in the survey were asked to estimate the numbers of Buzzards breeding in the areas chosen and defined by them. They were asked to estimate non-breeders and to provide notes on the previous history of the Buzzard in the census area. Helpers were also asked to make breeding records on the forms or on BTO nest-record cards, and to get in touch with me if they could help with a study on the feeding habits of Buzzards, either by direct observation or by collecting the pellets regurgitated by Buzzards.

There was an excellent response to the request for information: 15 171 km^2 (5857 square miles) were surveyed and it was estimated that 810 pairs bred in this area; 357 nests were actually found. The breeding records showed that the average clutch size was between two and three. Thus we obtained a fairly reliable picture of the situation which existed before the rabbit population had been greatly reduced. A smaller sample involving thirteen areas was repeated in 1955 and the whole 1954 sample was repeated in 1956, by which time myxomatosis covered practically the whole of Great Britain.

The clearest way to describe what happened is to record the changes in the thirteen areas which were studied in all 3 years. In 1954, when rabbits were still plentiful, 175 pairs bred, but in 1955, when most of the rabbits had gone, only 23 pairs bred. Despite the low numbers of rabbits in 1956, the number of pairs of Buzzards breeding increased to 58 pairs in that year. In each of the 13 census areas there was a decline in breeding pairs between 1954 and 1955. Between 1955 and 1956 declines were observed in two areas, the numbers breeding remained the same in six areas and they increased in five areas. It proved impossible to measure exactly the change in clutch size between 1954 and the later years since few nests were observed right through the season. Thus a report of one egg in a nest could either mean that only one egg was laid or that the observer happened to visit the nest when one egg was laid, but it was in fact the first of a clutch of two or three eggs. However, the information obtained suggested that the average clutch size when rabbits were unavailable was between one and two eggs whereas in 1954 it had been between two and three.

I was able to conclude that myxomatosis had had a significant indirect effect on the reproduction of the Buzzard. It had caused a reduction in clutch size in those that bred and a failure to breed in many pairs. However, there were clear signs of a recovery in 1956. This showed that the Buzzard, to some extent at least, was able to find alternative food. This was not surprising, since even when rabbits were still abundant in the spring of 1954, no fewer than sixty-four prey species were recorded in the survey. These included small mammals, birds, carrion, beetles, earthworms and even berries. No evidence was found to suggest that the range of the Buzzard had been reduced by myxomatosis.

Further surveys were planned but in 1960 it became obvious that I could not organise them as I had to switch my attention from the subject of biological control to that of pesticides. The breeding records obtained in the three surveys were used by Colin Tubbs in his admirable book on the Buzzard. It will be extremely interesting to compare the results of the 1983 BTO Buzzard survey organised by Dr Kenneth Taylor with those obtained in the 1950s.

In 1956 I was asked by the editors of *British Birds* to contribute a paper on the Buzzard in their series of papers on birds of prey. This provided me with an opportunity to report the findings of the 1954 survey in their historical context (Fig. 32). My studies showed that the species had originally been – as Pennant described in 1776 – 'the commonest of the hawk kind we have in England'. It was found throughout the British Isles. By the 1860s when A. G. More carried out his pioneer BTO-type surveys on the distribution of British breeding birds, the Buzzard had disappeared from much of the eastern part of Britain. In my paper I showed that in 1954 the highest population densities of breeding Buzzards occurred on farmland, but on farmland that was not keepered. Overall, the distribution of the Buzzard was negatively correlated with the population density of game keepers (Fig. 33). This showed that the Buzzard was restricted to the west and north of the country not because it preferred moorland but because it had been exterminated by game keepers in the east. I expressed the view that the future of the Buzzard would not depend on the rabbit but 'will depend, like its past, on the

Fig. 32. Changes in the distribution of the Buzzard (*Buteo buteo*) in the British Isles, 1800–1954. Black, breeding proved, or good circumstantial evidence of breeding; ? on black, circumstantial evidence suggests that breeding probably took place; ? on white, inadequate evidence of breeding; white, no evidence of breeding. (Moore, 1957. Reproduced by permission from the monthly magazine *British Birds*.)

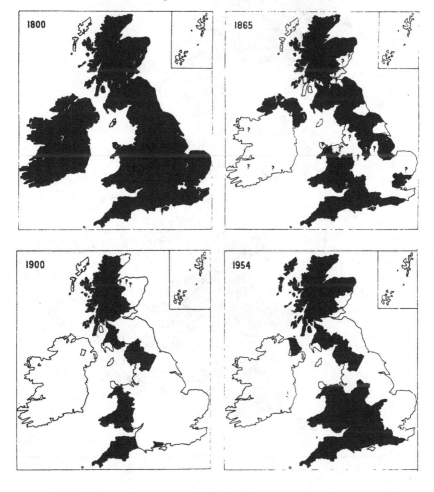

Fig. 33. The population density of Buzzards and game keepers. (*a*) Double cross-hatch, average breeding density of Buzzards equals 1 or more pairs per 10 square miles; cross-hatch, average breeding density of Buzzards is more than 1 pair per 100 square miles, but less than 1 pair per 10 square miles; diagonal hatch, average breeding density of Buzzards is less than 1 pair per 100 square miles; white, no breeding Buzzards. Data obtained in 1954.

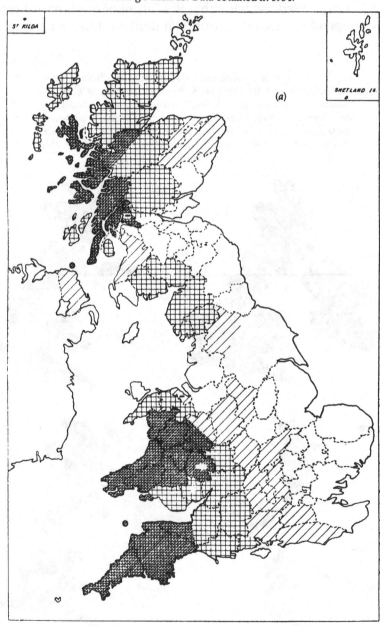

Fig. 33 *continued*. (*b*) Double cross-hatch, 3–6 members of The Game Keepers Association per 100 square miles in 1955; cross-hatch, 1–2 members; diagonal hatch, 1–0.5 members; white, fewer than 0.5 members; G principal grouse-preserving areas. (Moore, 1957. Reproduced by permission from the monthly magazine *British Birds*.)

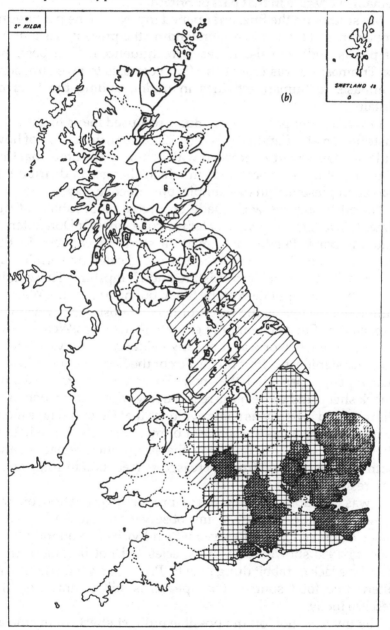

opinions of those who preserve game'. This view has been confirmed by subsequent events. Despite repeated attempts to breed in its old range in the east it is rarely successful and remains very rare in that part of Britain.

My studies on the Buzzard opened my eyes to the paramount importance of history in understanding the present distribution of plants and animals. I was also influenced, I suspect, by Dr Thomas; he was one of the first botanists to give sufficient emphasis to human activities in understanding the flora of Britain.

In writing my paper on the Buzzard I used maps to show its distribution at different times (Fig. 32). It is an indication of how little biologists were concerned with history at the time, that this was one of the first occasions in which changes in distribution had been presented in this simple, but effective way.

No other surveys were made on predators of rabbits at the time. However, many years later my colleagues Dr Don Jefferies and Mr Barry Pendlebury, made a study of changes in the populations of Stoats and Weazels as indicated by vermin bag records at estates in Hertfordshire and Hampshire. These indicated that the population of the Stoat, but not that of the Weazel, suffered a decline as the result of myxomatosis. This was to be expected as Stoats commonly eat rabbits whereas Weazels live mainly on mice. Nor was there any evidence that myxomatosis had any significant effect on the Fox or the Badger. Nevertheless many people assumed that both Foxes and Buzzards would attack sheep and game birds respectively when their rabbit prey disappeared. There was some evidence that Foxes did take more lambs in some areas at the onset of myxomatosis, but whether they did so significantly or not, the supposition remained and both species suffered from increased persecution in some areas as a result.

I was not able to carry out a proper investigation on the Minotaur beetle. However, my observations on the Hartland Moor National Nature Reserve which I visited frequently from 1953 to 1960 showed that the species did not become extinct despite a lack of rabbit dung pellets. Presumably it too found an alternative food source. The species is still common on the reserve today.

The investigation on the possible indirect effects of myxomatosis on the hare was undertaken at the same time as that on the

Buzzard. The hare was chosen as the representative of rabbit competitors because it was large and diurnal and therefore easy to observe and because there was some evidence that it competed in some way with the rabbit. In December 1953 Ronald Lockley undertook a survey of myxomatosis in France on behalf of the Nature Conservancy. His report included a section on the hare as a food competitor. Where rabbits had become very abundant, the hare had disappeared, but it made a rapid return in such areas when rabbits had been cleared out by myxomatosis. Some thought that the removal of rabbit fences had facilitated its return. Ronald Lockley was frequently told that rabbits attacked hares and that they did so by tearing open the scrotum of their enemies! No scientific evidence was ever produced in support of this belief. During the period of the hare investigation, I tried to find places where I could watch interaction between hares and rabbits but, owing largely to the rapid decline of the rabbit, I was unable to find any.

By 1955 a few authenticated cases of both Brown Hares and Mountain Hares being affected directly by myxomatosis were recorded. These occurrences were extremely rare and remained so. If there were any effects they were likely to be indirect.

It was obviously desirable to make accurate sample censuses of hares before and after the arrival of myxomatosis. The work of Dr J. Andersen in Denmark had shown that, despite its size, the hare was almost impossible to count accurately; my own observations supported this view. Therefore the only way we could measure the indirect effects of myxomatosis was to study game bag records and to see if hares invaded areas where they had not been seen in recent years. I sought help from the newly formed Mammal Society of the British Isles and from Pest Officers of the Ministry of Agriculture. The information obtained showed that the hare was found in varying numbers throughout most of England and Wales. It was only absent from urban areas, high mountains, an area of the Sussex Weald, southwest Wales and much of the western part of the Cornish/Devonian peninsula. There was good evidence to show that at least in the Weald area and Pembrokeshire the hare had once been common. As in France, it had apparently disappeared from these areas because of the great increase of rabbits during the period of agricultural depression. However, when myxomatosis killed off most of the rabbits in these areas they were not recolonised by hares on a large scale. In some more localised areas hares did return and an increase in hares was reported on a number of shoots in 1955. However, it was not certain that this was due to the lack of rabbits; 1954 was a bad year for hares owing to bad weather and disease and so an increase in 1955 was to be expected in any case. Many people reported more hares in woodland than hitherto; it seems that the increase of grass and other food in woods due to the absence of rabbits may have been the cause.

Thus it seemed that the hare was helped slightly by the decline of its competitor. However, it is a species where populations are known to fluctuate very considerably. Since these natural fluctuations exceeded those due to myxomatosis it was not possible to quantify the latter. The results of the hare investigation were disappointing scientifically, but I believe we were right in using the unique opportunity to try to study competition between two common species on the grand scale. One of the admirable features of the study was that it showed that agricultural and conservation biologists could work together on a problem that concerned them both.

No special surveys were made on the indirect effects of myxomatosis on the Wheatear and the Stone Curlew. Both were declining due to the loss of grassland to forestry and, in the case of the Wheatear, of grassland to arable. It is probable that myxomatosis accelerated the declines but the extent to which this occurred could not be determined by general observation.

No studies were planned or made on the one species, which (with the benefit of hindsight) we can now be fairly certain was

Fig. 34. Large Blue (*Maculinea arion*). The extinction of this species in England was largely due to myxomatosis. Reduction in grazing by rabbits eliminated the thyme and the ant on which it depended for its survival.

indirectly exterminated by myxomatosis. The species is the remarkable and beautiful Large Blue Butterfly (Fig. 34). I only once saw them myself. They were feeding and flying on the south-facing side of a sheltered valley which ran down to the sea on the north Devon coast. I was struck by their size – at first sight they seemed more like a white than a blue butterfly. They flew strongly in the hot sun.

The life history of this species is extraordinary. Its larva is dependent on thyme, on which it feeds until its last larval stage or instar. It is then picked up by an ant *Myrmica subuleti* which carries it to its nest. There it feeds on the ants' offspring, paying for the ants' hospitality by producing a secretion on which the ant feeds. The Large Blue larvae pupate in the ants' nest before emerging. Thus, this insect is dependent on both thyme and a rather uncommon species of ant, and they are dependent on grazing which keeps the turf short. The grassland on the edges of cliffs was a favourite haunt of the butterfly. When these were fenced off by farmers, only rabbits could keep the turf short enough, and when the rabbits were killed off by myxomatosis the vegetation grew up and both thyme and ant disappeared and the Large Blue with them.

Two reserves were set up to protect the butterfly, but neither succeeded in their purpose. In one, the grazing pressure was insufficient to maintain the short turf. In the other, the population of the butterfly was too small to be viable and, despite the provision of an excellent grazing regime based on the researches of Jeremy Thomas, it became extinct following a bad season in 1979.

Ploughing up and reseeding grassland greatly reduced the habitat of the Large Blue, but myxomatosis almost certainly gave it the *coup de grâce* by removing indirectly its remaining food supply.

In conclusion, we can say that the indirect effects of myxomatosis on vertebrates appear to have been relatively slight. Those on plants (and hence insects) were probably much greater, but they were masked by the much greater changes due to changes in grassland husbandry and forestry practice which were occurring at the same time. Looking back on this period the most interesting aspect of myxomatosis was the reaction of people and government to an unusually effective method of biological control.

11

*

The politics of myxomatosis

The use of myxomatosis to control rabbits in Australia and France caused little concern in Britain but, when the disease was introduced into this country, it released a torrent of opinion and argument in the media and in ordinary conversation. For the first time a biological problem held the interest of the general public for months on end and forced the government to intervene.

A perusal of the numerous letters written to *The Times* in October and November 1953 showed that the country was sharply divided between those in favour of myxomatosis and those against it. Those in favour were mainly farmers and foresters, who were aware of the immense damage done every year by rabbits. They saw the arrival of myxomatosis as a heaven-sent opportunity to get rid of a very serious pest. The more scientifically minded realised that it would be virtually impossible to exterminate the rabbit, but that its reduction would nevertheless bring great benefit to farming and forestry.

The annual damage caused by rabbits was estimated at £50 000 000. However, rabbits were also a valuable source of food and their fur was used for hats and other clothing. Their value was estimated at £15 000 000. Not surprisingly, those who earned their living by catching rabbits and those in trades dependent on rabbits were very much against myxomatosis. Its opponents included sportsmen, especially those addicted to rough shooting, who saw that they would lose one of their most abundant quarries. However, the most vocal were the general public who were horrified by the sufferings of diseased rabbits.

The general public were not affected directly by loss of income due to rabbits and their image of the species was doubtless more influenced by Beatrix Potter than by farmers. So there was little sympathy for the farmers' views on rabbits. Only farmers who enjoyed a little bit of rough shooting felt the full force of the dilemma. As the writer of a long and balanced article in *The Times* of 11 November 1953, entitled 'Pestilence and pest: rabbits and the "blind death" from France', remarked, 'Whatever course the disease runs in Great Britain the Ministry of Agriculture has an unenviable task'. On 23 July 1954, the leading article in the *Shooting Times* entitled 'Witness this Myxomatosis' ended with these words 'But if good ever does come out of this evil, it may take the form of proper, reasonable, organised control of the rabbit – and nothing but contempt for those farmers who today are artificially spreading, or causing to be spread, a form of beastliness altogether foreign to the ideas of any decent Britisher'. The problem was not as simple as that because myxomatosis was much the most efficient way of controlling rabbits so far devised.

Considering that the disease had first been confirmed in this country on 13 October 1953 the government acted with commendable speed. The Minister of Agriculture, Sir Thomas Dugdale, announced the setting up of the Advisory Committee on Myxomatosis that same month, and the Committee held its first meeting on 4 November 1953. The Committee was chaired by Lord Carrington, who was then the Joint Parliamentary Secretary to the Minister of Agriculture and Fisheries, and its distinguished membership included the Hon. Miriam Rothschild, the expert on fleas, and Major R. B. (later Sir Ralph) Verney who was later to become the Chairman of the Nature Conservancy Council. The Committee set up a scientific sub-committee to advise it on the scientific aspects of the disease and this sub-committee was assisted by a small research group. As mentioned above, I was appointed to the scientific sub-committee on the instigation of Max Nicholson. By 16 March 1954 the Advisory Committee on Myxomatosis had held six meetings and published its first report.

The first problem which had to be tackled was whether to try to prevent the disease spreading from the three places in which it occurred. When the first outbreaks of myxomatosis were reported at Edenbridge in Kent and Robertsbridge in East Sussex, the Ministry of Agriculture tried to contain them by surrounding the affected areas with rabbit fencing. These measures failed in their purpose. The third outbreak, which occurred on the edges of the Lewes-to-Eastbourne road, could not even be fenced. Thus the Advisory Committee were faced with a *fait accompli*: it was virtually impossible to prevent the spread of the disease. As a result they recommended that no attempt should be made to stop it.

Their second problem was much more difficult. Should they recommend the government to assist the spread of the disease? From the farming point of view the answer was 'yes', but public opinion would have been outraged if the government had supported the spreading of something so horrible. Nevertheless I personally believed that we should disseminate the disease on humanitarian grounds; if we allowed it to spread naturally it would remain in the country for many years and far more rabbits would die of it than if the population were reduced at once by using the disease itself to achieve this aim. I discussed this with Lord Merthyr, who represented the Royal Society for the Prevention of Cruelty to Animals (RSPCA) on the Advisory Committee. He agreed with my logic, but said that he could never convince his members that it was right to use the disease at all, so this approach (which several advocated) was dropped.

Meanwhile it was becoming increasingly obvious that farmers were taking the law into their own hands and introducing diseased rabbits on to their land in order to infect their own rabbits with myxomatosis. This generated strong public pressure to make such actions illegal, and the Advisory Committee, aware of the strength of the humanitarian argument, formally recommended that no attempt should be made to assist the spread of the disease in this way. The politicians bowed to public pressure and the 1954 Pests Act included a section which specifically prohibited the use of a rabbit with myxomatosis to spread the disease among uninfected rabbits. We were told that this section received special support from the Prime Minister, Sir Winston Churchill, who was well aware of its political value. Even so, proposals to repeal this very section were made the same year, which suggested that the farmers' views were gaining ground.

Two other matters had to be considered by the Advisory Committee. Domestic rabbits could catch myxomatosis from wild rabbits in those few areas where mosquitoes as well as fleas transmitted the disease. Accordingly the Advisory Committee advised the government to promote research into finding suitable vaccines which could be used to protect domestic rabbits. The other problem concerned the proposal, made in both Britain and France, that Cottontail rabbits (Fig. 35) should be introduced from America to provide a quarry in place of the European rabbit. The suggestion was based on the knowledge that Cottontails were not killed by myxomatosis. The dangers of introducing yet another pest species into Britain were obvious and so the Committee advised that action should be taken to prevent it and no Cottontails were introduced.

Realising that the spread of myxomatosis (both naturally and by human agency) could not be stopped, the public learnt to live with the distressing sight of dead and dying rabbits in every corner of the land. The animals did not cry out and they continued to feed until they died. People comforted themselves by observing that the death they suffered was no worse than that in the jaws of a Fox. Eventually a virtue was made of a necessity. On 13 October 1955, the second anniversary of the introduction of myxomatosis, the Ministry of Agriculture told a press conference 'I think it is no exaggeration to say that as a result of

Fig. 35. Eastern Cottontail (*Sylvilagus floridanus*). Myxomatosis is endemic in this American rabbit, and does it little harm. The Myxomatosis Advisory Committee was successful in opposing the proposal to introduce this species as an alternative quarry to the European rabbit, when the latter succumbed to myxomatosis.

myxomatosis we may well be approaching a new phase in British agriculture and I feel that the press, as mirrors of history, should be in on it'. Indeed, the second (1955) report of the Advisory Committee on Myxomatosis, now chaired by Lord St Aldwyn (Lord Carrington had become Parliamentary Secretary to the Ministry of Defence), was mainly concerned with the problem of how to provide help to 'occupiers in eliminating rabbits that survive outbreaks of myxomatosis'. The 1954 Pests Act had made it possible to set up clearance areas for rabbits. If owners did not clear rabbits from these areas, the Ministry of Agriculture was empowered to do so. In 1957 the Advisory Committee on Myxomatosis was wound up and an Advisory Council on Rabbit Clearance instituted in its place.

Throughout the political debate on myxomatosis the scientific aspects of the problem were widely discussed and research on it was promoted. In 1955 the Mammal Society of the British Isles held a symposium on the subject, and in 1956 there was a special session on myxomatosis at the 6th Technical meeting of the International Union for the Conservation of Nature and Natural Resources (IUCN) which was held in Edinburgh. At that meeting, representatives of all the main countries affected by myxomatosis described their experience with the disease. The situation in Britain did not seem to differ greatly from that in other European countries.

The research of Harry Thompson and Mr Charles Armour of the Infestation Control Division of MAFF recorded in detail the spread of the disease from the original outbreak at Edenbridge. The work of Ronald Lockley, Miriam Rothschild and R. C. Muirhead Thompson demonstrated that the flea was the principal vector of the disease in Britain, the mosquito playing only a very minor role. Meanwhile research on the use of fibroma and myxoma viruses was done by Dr W. Mansi and others in support of the inoculation programme for domestic rabbits. As recorded in the last chapter, Dr A. S. Thomas and I were involved with studies on the indirect ecological effects of myxomatosis.

It must be admitted that all this research had little or no effect on the spread of the disease or on the government's attitude to it. However, the general opinions of biologists were probably valuable to the government and to the public. We could predict that the virus would become attenuated and that the rabbit would become resistant, therefore the country could not rely on myxomatosis controlling rabbits indefinitely. We could predict that the decline of the rabbit would have numerous effects, even if we could not predict their exact nature. As things turned out, myxomatosis did not result in the extinction of any plant or animal, with the probable exception of the Large Blue Butterfly.

From an academic point of view, the advent of myxomatosis was an extremely interesting one. The sheer scale of the effects was notable and must have made many biologists even more aware of the potential danger of introducing non-native species to this country. I believe also that the magnitude of the problem forced conservationists to take a wider view of their profession. Certainly, in my case, it prepared me for the next great challenge to wildlife: that of pesticides.

To the general public, myxomatosis was not seen as an example of biological control; when, a few years later, biological control was extolled as a substitute for the use of pesticides few people were restrained by their experience of myxomatosis. The public's main concern was for the rabbit and for themselves. They were outraged by the horrible appearance of the disease, and worried that circumstances might arise in which they could be poisoned by eating diseased rabbits or in which they could get the disease themselves. Subsequent experience has shown that myxomatosis is a remarkably specific disease – apart from rabbits, only a few hares have caught it. However, the myxoma virus is related to the pox viruses, some of which are deadly to man. The fear that there might be a mutation in myxoma which made it dangerous to man was a reasonable one. Today, when there is increasing interest in viruses to control insect pests, those who operate the Pesticide Safety Precautions Scheme (which includes biological agents as well as chemical ones) would never give clearance to any virus belonging to a group which affected vertebrate animals. As we shall see in the next chapter, the use of pesticides raises immense problems for conservationists but at least pesticides cannot mutate.

12

*

Pesticides – a new problem

The following chapters describe the problem for conservation posed by the extensive use of chemicals to control pests, diseases and weeds during the post-war period. I shall recall how research was developed at Monks Wood in order to understand how pesticides were affecting the environment. I shall then record how the results of that research were used to obtain better control of pesticides, and how they influenced the control of other pollutants. I shall conclude by describing how research on the effects of pesticides influenced the environmental movement as a whole.

Pesticides are chemicals used by man to kill plants and animals. The word pesticides is a useful general term which covers herbicides, fungicides, insecticides, molluscicides, rat poisons and similar chemical agents used in preventive medicine, agriculture, horticulture and forestry. It is a new word – it is not mentioned in the 1950 edition of the *Concise Oxford Dictionary* and even today its common use has not entirely settled down, for it is still sometimes used in a narrow sense (as a synonym of insecticides). The word pesticide covers natural substances, like pyrethrum, as well as the products of chemical industry; its meaning is stretched to include compounds – like maleic hydrazide – which regulate the growth of plants but do not kill them.

Thus defined, pesticides may seem a rather dull and technical topic, but it is quite otherwise. The rapid development of pesticides in the mid-twentieth century raised entirely new problems for nature and society. Pesticides became symbolic of the agricultural revolution of which they were an important part, and they more than anything else engendered a fundamental shift in our attitude to the environment. This extraordinary event occurred within the span of my working life. What follows is not a comprehensive history, but an account by a biologist who observed the events at close hand, and who endeavoured to reduce some of the harmful effects of pesticides on wildlife.

The chemical revolution of agriculture had a long gestation period. Salt and ashes have been used to keep down weeds for centuries. Nicotine was first applied as an insecticide in 1763. Sulphur was used against powdery mildew of the vine in 1848, and Bordeaux mixture against downy mildew of the vine in 1882, Paris Green against Colorado Beetle in 1867, and lead arsenate against the same pest in 1892. That year also saw the first marketing of dinitrocresol, the first of the synthetic organic pesticides. Further developments, notably on organomercury fungicides, occurred in the earlier part of the present century. Numerous scientists and agriculturalists advocated the use of chemicals, both pesticides and fertilisers, in the 1920s and 1930s, but agriculture was so depressed in those decades that farmers made little use of the new discoveries; labour was cheap and farmyard manure was available. Agricultural theory was of the twentieth century but farming practice largely of the nineteenth. It needed the Second World War to bring the two together and produce the chemical revolution of agriculture. So long as pesticides were used on a small scale their medical and environmental effects were trivial, but when they began to be used extensively side effects became more obvious and demanded attention.

Ecological repercussions were to be expected. Myxomatosis had shown how just one control agent, which affected only one species, could have complicated and largely unpredictable effects on numerous other species. No chemical firm has yet been able to produce a pesticide which is specific, that is it kills the pest and nothing else. To design such a pesticide requires a depth of knowledge about the physiology of individual species which we do not yet possess. When a pesticide is used against a pest it always has direct effects on some other species so it is likely to have even more complicated effects than myxomatosis which affects only one species directly. The likelihood of damage by pesticides seems obvious today, but in the early years of pesticide use, the benefits to human health and food production conferred by DDT, BHC and the growth-regulating hormone weed killers were so great that few people bothered to consider the problems which their use would inevitably produce. Several years passed before ecologists began to look at pesticides from an ecological point of view. Yet it was, and still is, crucial to understand their effects in ecological terms. This becomes clear when we see how they impinge on the environment.

Animal populations are affected by changes in the death rate and the birth rate. Pesticides can affect both directly by poisoning, and indirectly by poisoning other organisms which are ecologically important to them as food, competitors or predators. Animals obtain pesticides directly by feeding on contaminated food, by drinking contaminated water or through respiration. Contaminated prey may be dead, moribund or alive. As we shall see, the ability of prey species to store some pesticides in their bodies without damage to themselves can have serious implications for the predators which feed on them.

The direct effects of pesticides are best described in toxicological terms. Any animal (including man) can be affected by any chemical – whether it does so or not depends on the dose rate. A poisonous chemical only differs from a non-poisonous one in that a smaller dose of it causes death or injury than in the case of a non-poisonous one. The commonly held belief that chemicals can be divided fundamentally into poisonous and non-poisonous ones is thus misleading. Our bodies are very good at dealing with small amounts of very toxic compounds such as the alkaloids in potatoes or with large amounts of much less toxic compounds such as sugar. However, species differ greatly in their reaction to the same chemical. What is safe for one animal is damaging for another. Thus DDT is rarely hazardous to man but is deadly to insects. At certain dose rates, pesticides may not kill adult animals but they may affect their reproduction, with the result that the animals produce fewer young than they otherwise would. These sublethal effects, like lethal ones, vary between species and between pesticides. Therefore, when a pesticide is applied to a crop it will usually kill a large proportion of the pest species against which it is used; it may also impair reproduction in the survivors. It will, furthermore, have variable, direct toxicological effects on the other species present in the same environment – it will kill some, have sublethal effects on some, and leave others unscathed. Thus the total direct effects of a pesticide application can be extremely complicated; the indirect effects resulting from the toxicological ones add yet further to the complexity. No species lives in a vacuum: each species in an ecosystem always depends on others, and it in turn affects other species to varying degrees.

For example, if a herbicide greatly reduces the population of a plant, all the herbivorous animals which feed on that plant will be affected: either they will have to feed on another plant or, if that is not available, they will starve. Similarly, if a prey species is greatly reduced by an insecticide, its predators will have to feed on other animals or starve. In general, the more specialised herbivores or predators will suffer more than those with more generalised habits. Pesticides do not have harmful effects on all the species in the sprayed environment. If either the predators or the competitors of a species are reduced by the application of a pesticide, the pesticide will actually favour that species. For example, the worldwide use of DDT against Codlin moths and other orchard pests killed off the predaceous mites and bugs which had previously held Red Spider Mites (Fig. 36) in check. As a result Red Spider Mites have themselves become a major pest throughout the world. The successful control of broad-leaf weeds such as Charlock and poppy in cereals allowed Wild Oats and Black Grass to increase greatly and become a serious economic problem. This has only been solved by the invention

Fig. 36. Red Spider Mite (*Metatetranychus ulmi*). This animal has become a pest in orchards in many parts of the world as the result of DDT killing the predators which used to keep it in check (much enlarged).

of sophisticated herbicides which kill these grass weeds, but not the closely related cereals.

Some of the indirect effects of pesticides involve more than one step. For example, the extensive use of herbicides to control cereal weeds has greatly reduced the insects which were dependent on those weeds. In turn this has reduced Partridge populations, because the Partridge chicks after hatching depend on an abundant supply of insects for their supply of protein, and the insects depended on the weeds.

Pesticides exert very strong selection pressures on the species affected: any individuals with genetic characteristics which makes them resistant to the spray will give them immense advantages. Thus resistance to a pesticide frequently develops in the more abundant species. Since the most abundant species are usually pests, this means that pests frequently become resistant to pesticides, and they do so before other, less abundant species can become resistant in the same environment. So the development of resistance yet further complicates the effects that pesticides can have on an ecosystem.

It is obvious that the effects of any one pesticide on any ecosystem, whether it is an agricultural or a forestry crop, or a wetland harbouring disease-carrying mosquitoes, must be immensely complicated. Real life situations are even more complicated because several pesticides are usually applied to the same bit of ground. For example, most corn fields are sprayed with herbicides. The corn seed is treated with a fungicide and an insecticide and, later, other fungicides are used to control rusts etc. and other insecticides to control cereal aphids. Orchards may receive ten treatments or more with different pesticides in one season. Treatments will often vary from year to year so that the plants and animals affected will be confronted with different pesticide situations in different years. The variations due to pesticides will interact with the vagaries of weather.

Enough has been described to show that the complexity of the effects of pesticides is so great that it is extremely difficult to study them, and so it is difficult to predict their effects. Enormous changes in the flora and fauna of Britain must have occurred in recent years due to pesticides but, in most cases, too little is known about populations before pesticides were used (and about their status today) for the effects to be quantifiable. Even when we can give some estimate of the changes, it is extremely rarely that we can describe the causes exactly, whether they are due to pesticides or not. This is the challenging biological background to the work described in the following chapters. Applied biologists, unlike academic biologists, cannot avoid problems just because they are difficult. The challenge provided by pesticides is formidable because the values and potential hazards of pesticides are both so great. Much work has to be done in order to get the maximum advantage from their use with the minimum of danger to man, domestic animals and the environment. Many problems caused by pesticides remain to be solved, but we can learn something about some of them by recalling our experience with the first ones which we encountered.

13
*

Planning research on pesticides

There is a special interest attached to the beginnings of events and organisations, to the times before ideas have become fixed and before wisdom has become conventional. One asks 'what was it like at the time?' In this chapter I shall try to describe how it felt when the pesticide problem began to impinge upon us and the need for better control of pesticides became apparent. I can best do this by describing my own involvement with pesticides.

I first encountered them as a student. In 1942 I was attending a Long Vacation course in zoology at Cambridge, before being called up for wartime service. We were taken to a demonstration of spraying equipment on a nearby farm. The chemical used was nicotine, and I remember being much more impressed by its potential dangers to the operator than by its potential dangers to the environment. My next encounter was more personal. I had spent the first three months of 1945 in a prisoner of war camp in Germany. Lack of food and medicine ensured that we had become the unwilling hosts of the body louse (*Pediculus humanis*). Before returning to polite society we had to be deloused. In an environment in which typhus was active, I was delighted to be liberally sprinkled with DDT, the wonderful substance which had been synthesised as long ago as 1874, but whose insecticidal qualities had only recently been discovered and applied.

I returned to civilian life and biology in 1947. I was deeply interested in wildlife and spent much of my time in the country during the next 5 years, yet it never occurred to me then that the increasing use of pesticides would have profound effects on the environment around me. It is difficult to see something if you do not look for it and I, like most naturalists and biologists at that time, was not looking for the effects of pesticides. However, they had been observed by some; the very toxic organophosphorus insecticides Schradan and Parathion, derived from war gases had been used to control cabbage aphid and had caused bird casualties in brassicas. An increasing number of game preservers and game keepers began to fear for their partridges and pheasants. Mr Wentworth Day even wrote a book in 1957 entitled *Poison on the Land: the War on Wildlife and some Remedies.* The main concern was naturally for human beings. A government committee had been set up under the chairmanship of Lord Zuckerman to consider what safety measures should be undertaken. Their first report was on pesticide residues in food; this was published in 1951. The second report (in 1953) recommended the formation of an interdepartmental Advisory Committee on Poisonous Substances, and this was quickly done. So little was known about the environmental consequences of pesticide use that the Zuckerman committee deferred pronouncements on them until special field studies had been made for the committee's perusal.

In the early 1950s, complaints had also been received about dinitrocresol (DNOC). This pesticide was widely used at that time both as a herbicide and an insecticide. Reports that hares and game birds had been turned yellow and had been killed by DNOC were quite widespread. The Ministry of Agriculture sought the help of the Nature Conservancy, and two of its recently appointed Regional Officers (Dr Eric Duffey from East Anglia and myself from the South West) were instructed to help with field trials in Norfolk and Berkshire respectively. In May 1954 I took part in a study of bird and mammal populations on areas treated with DNOC and on untreated controls. We tried to measure the effects of spraying by making observations before and after spraying. It was my first encounter with the problems of doing research on the effects of pesticides. I learnt how extraordinarily difficult it was to obtain results. We had only a few observers, and that meant we could cover only a relatively small area of farmland. It soon became obvious that differences in populations due to movements of birds and mammals about the farm could mask any differences due to poisoning, unless these were catastrophic in their effect. Pheasants, partridges and hares are large animals, but they are not nearly as easy to count as one might assume – we became increasingly suspicious about the accuracy of our counts. It was all too obvious that this type of work could not be done on the cheap and, with the limited resources available we could not provide very useful information for the Zuckerman Committee. We simply could not measure the effects of DNOC spraying accurately; all we could say was that their effects did not seem to be disastrous in the areas in which we had worked.

In the 1950s the Regional Officers of the South and South West Regions were based on Furzebrook Research Station near Wareham. My colleague from the South Region was Miss Olive Balme (now Mrs Young). She had also been involved with a pesticide problem but of a different kind. She and Dr Arthur Willis (now Professor Willis) had been studying the effects of selective herbicides on the flora of roadside verges in Gloucestershire. Their work showed that profound changes in the flora of the verges were caused by the spraying. At that time, Gloucestershire was unusual in using weedkillers to control roadside verge vegetation. The situation was potentially dangerous from the conservation point of view but the operations were on too small a scale to be nationally significant. However, it was not long before other counties began to use herbicides on roadside verges, and the seriousness of the threat to the flora and to the appearance of the countryside was perceived. The Nature Conservancy made strong representations to the Ministry of Transport, who issued Circular 718 in 1955. This recommended strongly that herbicides should be used only to control noxious weeds and only in places where road safety demanded it. The circular was successful in preventing most County Councils from using herbicides on their verges indiscriminately.

In the same year the Advisory Committee on Poisonous Substances was enlarged in order to take account of environmental issues. The Nature Conservancy was first represented by Mr R. E. Boote, its Establishments Officer, and Olive Balme served on its scientific sub-committee. Pesticides were clearly becoming a problem, but at that time everything appeared to be reasonably under control – not for long. In 1954 – unbeknown to us – the first field trials on two very toxic persistent organo-chlorine insecticides were being carried out by Shell. The first vertebrate wildlife casualties from these chemicals were reported in 1956. The reports increased and by 1958 the Nature Conservancy, and notably Bob Boote, perceived the need for the Nature Conservancy to take an active role in reducing pesticide hazards. In 1959 the Huntingdon working party under the chairmanship of Dr I. Thomas of MAFF was set up to advise on the location of what was to become Monks Wood Experimental Station, which was to be the base for the Nature Conservancy's research on pesticide effects. I served on the working party, having been asked to take charge of the pesticide studies. I relinquished my post as Regional Officer in 1960 and began to plan the pro-gramme of the Toxic Chemicals and Wildlife Section (TCWLS).

When painting a picture there is something peculiarly intimidating about the completely blank canvas which stands before you. I had rather similar feelings when confronted with designing the research programme of the Toxic Chemicals and Wildlife Section. The possibilities seemed infinite: dozens of different pesticides were already on the market and thousands of plant and animal species might be affected by them. All we knew for certain was that a wide range of pesticides sometimes killed or damaged a wide range of organisms. The only obvious effects were herbicide drift, which caused browning and distor-tion of hedgerow plants when wind blew herbicides into hedges, and dead birds, which were sometimes found in fields after they had been sprayed with insecticides.

The purpose of the work of the Toxic Chemicals and Wildlife Section was to determine which pesticide effects were important so that we could give advice on how to reduce serious hazards to wildlife. Toxicological studies in the laboratory could determine whether individuals of a given species were likely to be affected by a given pesticide in the field. Only extensive observations in the field could indicate whether a pesticide was actually causing casualties; only well-designed field experiments could give com-plete proof that a given pesticide did or did not have a significant effect on a population. If we were to carry out observations and experiments in the field we had to have a baseline. In other words, we had to know what was there before we made any observations on pesticide effects or carried out any experiments.

My initial studies of the literature showed that baseline studies did not exist; next to nothing was known about wildlife on ordinary farmland. As noted earlier (p. 44), work on hedges had been virtually confined to the distribution of Spindle and its role as a host to the Bean Aphid (*Aphis fabae*). Apart from the pioneer work of Mr W. B. Alexander, who had studied the birds on an Oxfordshire farm before the war, practically nothing had been done on the relationship between birds, crops and hedges.

My experience of the Zuckerman Committee trials in 1954 had shown me that work on pesticides required control of quite large areas of land, and that the work was labour intensive. I realised that the effectiveness of our work would depend largely on the extent to which we could use experimental land and on the numbers of people we could employ on it.

I had to plan the details of our programme within the likely constraints of land and manpower. Numerous decisions had to be made. Should we concentrate our work on species like birds, which were easily observed, or on invertebrates, whose populations could be more easily studied experimentally in the field? Should we study rare species or common species? Which species were amenable to toxicological research at Monks Wood? Clearly we should concentrate on those pesticides which posed the greatest threat to wildlife, but were they the most widely used ones or the most toxic? Baseline studies on the wildlife of farmland were obviously necessary, but what proportion of our time should be spent on them compared to the studies of actual pesticide effects?

These were all difficult questions, and I was fortunate in having plenty of time to brood over them in 1960 as I finished off my studies on changes in the Dorset heathlands. I was able to talk to numerous biologists, both academic and applied; by the end of the year, I had no doubts about what the basic pattern of our work should be. We should study the whole system, the background of the pesticide effects as well as the effects themselves, the natural history of hedges as well as the toxicological effects of pesticides on species living in them.

I decided we could and should start work on the background of pesticide effects straight away as that did not require the use of laboratories. Since so much of the wildlife on the farm was concentrated in the hedges, we first had to study different sorts of hedge and the plants and animals which they contained.

The choice of what type of pesticide should be studied was more difficult. There was no doubt that selective herbicides like MCPA were the most widely used – clearly they had immense effects on the weed flora and, presumably, on animals which depended on the weeds; but even if we could have measured their effects, what value would that have held for conservation? The herbicides were only doing the job of the hoe, but more effectively. The Nature Conservancy could not ask farmers to weed less efficiently in order to protect weeds.

The insecticides which appeared to be killing the most birds and mammals were very toxic organophosphorus substances such as Schradan and Parathion and the new organochlorine insecticides aldrin, dieldrin, endrin and heptachlor. In 1960, Mr Charles Elton drew my attention to a recently published paper in a little known journal in California. The paper, entitled 'Inimical effects on wildlife of periodic DDD applications to Clear Lake', had been written by Mr E. G. Hunt and Mr A. I. Bischoff, both of whom worked for the Californian Fisheries and Game Department. The paper showed how an organochlorine insecticide of low toxicity had had a major effect on the population of Western Grebes of Clear Lake (Fig. 37). The chemical, now more usually known as TDE, had been applied by aircraft to control a gnat *Chaoborus astictopus* which was causing a nuisance to those living by the lake. Hunt and Bischoff carried out chemical analyses of the pesticide in plankton, fishes of different feeding habits and the Grebes. This work showed that the

Fig. 37. Western Grebe (*Aechmophorus occidentalis*). This fish-eating bird is at the top of the food chain on Clear Lake (see Fig. 38). Its population was greatly reduced by consuming fish contaminated with the persistent organochlorine insecticide TDE, which had been applied to control gnats.

persistence of a pesticide was an extremely important factor and its effects could not be predicted easily. These facts convinced me that, in the long run, persistence could be more important than toxicity. It mattered more that pesticides lasted a long time in the environment than that they were very poisonous. So I decided that we should give priority to those pesticides which were known to be persistent – not only the very toxic ones such as aldrin and dieldrin, but also the far less toxic ones DDT, TDE and BHC.

So the strange events at a lake thousands of miles away in California had a large share in determining pesticide/wildlife research in Britain. Of course, in subsequent years, the Clear Lake story became well known through the writings of Rachel Carson. It was the food chain effect that everyone had heard about. As we shall see, the situation was not as simple as at first believed, yet the story contained much truth and it remained a potent myth.

Some years later I was staying with Professor Robert Rudd who had done so much in the USA to warn people about the hazards of pesticides and to promote research on their ecological effects. Bob Rudd knew of my great interest in Clear Lake and kindly arranged for us to see it both from the water and the air (Fig. 38). As we flew round it in a light aircraft we could see the numerous holiday homes along its shores (whose occupants had complained about the insects which had necessitated the spraying). I could see there had been a real problem. When we explored the marshy shores of the lake in a boat, we got close to some of the surviving Western Grebes. The sight of them made me all the more determined to get the unnecessary use of persistent insecticides outlawed and to seek alternatives that did less damage to wildlife.

From the point of view of conservation, most pesticides were likely to be damaging rather than beneficial. It would have been easy for our section to develop an anti-pesticide stance which would have served it ill when it was trying to persuade agriculturalists to modify their procedures in order to reduce the environmental hazards of pesticide use. I was therefore keen to develop a subsidiary programme in which we could demonstrate their value to conservation as well as their dangers. One of the most notable and damaging side effects of myxomatosis was the excessive growth of hawthorn scrub on grasslands. The scrub was difficult and expensive to control physically, but we felt it might be possible to control it cheaply and effectively by using herbicides. One of the main problems in scrub control is that when the bushes have been cut down new growth shoots up from the stumps in the following spring. This can be prevented by applying an appropriate herbicide on the cut stump in order to kill it. If we could show that this method was both effective and safe, it would provide a useful tool and would demonstrate that pesticides had their uses for conservation as well as providing hazards for wildlife. Eventually our experiments on scrub at Bookham Common in Surrey, Wicken Fen in Cambridgeshire and High Halstow in Kent showed that stump treatment with appropriate herbicides was effective and it caused no damage to the surrounding vegetation or to the soil animals living among the roots.

However, our immediate task was to convince the Nature Conservancy committees that our programme should include such studies at all. Our proposals for baseline studies, for

Fig. 38. Clear Lake, California. It was here that Hunt and Bischoff first demonstrated the food chain effect of persistent organochlorine insecticides.

studies on the effects of persistent insecticides and for studies on the use of pesticides as conservation tools were all put to the Nature Conservancy in 1961 and were accepted. They were amplified in the following years and programmes on other topics were added, but they remained the basis of the work of the TCWLS throughout its existence. Some of the background work on hedges has already been described (see Chapter 4). The next chapters deal with work on the persistent organochlorine insecticides, which were our main concern.

14

*

Persistence

When a farmer uses a pesticide he usually wants it to disappear quickly so that it does not leave a residue on the crop which could poison people or animals which eat it. Therefore most pesticides used by farmers are those which break down fairly rapidly into harmless constituents. However, persistent pesticides do have their uses both in industry and in agriculture; as some of them have serious effects on wildlife, it is worth considering persistence in some detail. Whether it is ecologically significant or not depends on the form of the pesticide which persists and on the extent to which it can disperse in the environment. Simazine, paraquat and DDT are all examples of persistent pesticides, but only DDT has significant ecological effects.

If vegetation is allowed to grow round certain industrial installations it can provide a fire risk in dry weather. Therefore there is a need for a herbicide which kills all the weeds round the installation and keeps the area clear of them for as long as possible. Similarly, gardeners like to keep gravel paths free of weeds for as long as possible after spraying them with a herbicide. Simazine is one of the chemicals which is suitable for these purposes because it kills almost all weeds and persists in the soil for many months as an active herbicide. It does not move in the soil and so its effects are restricted to the target area.

Farmers often want to substitute productive strains of grass for mixed swards of less productive species without having to plough the land. They may also wish to rid a field of weeds and drill corn in it without ploughing. The herbicide paraquat is suitable for these purposes because it kills the weeds when applied but, on contact with the soil it becomes inactivated and so grass or corn can be drilled immediately after spraying. Although the paraquat is inactivated it does not disappear – its molecules are held in the soil lattices. Paraquat is highly persistent but, since it is not released from the soil, its persistence has no harmful ecological effects.

The seed of corn, sugar beet and other crops is often attacked by fungal and insect pests. Wheat Bulb Fly attacks wheat seed many weeks after sowing and is a particularly serious pest. Therefore there is a great need for persistent fungicides and insecticides which can give protection to the seed from sowing to germination. In fact, most cereal and all sugar beet seed is treated with a fungicide, an insecticide or both. In the 1950s and 1960s nearly all the insecticides used as seed dressings belonged to the group known as organochlorine or chlorinated hydrocarbon insecticides. Some of the insecticides in the group, such as dieldrin, are themselves persistent. Other organochlorine insecticides are not persistent themselves, but their metabolites or breakdown products are. For example, the insecticides heptachlor and aldrin are much less persistent than their respective metabolites heptachlor epoxide and dieldrin. Most DDT is turned into DDE which is extremely persistent.

Unlike other persistent pesticides, the organochlorine insecticides are very soluble in fat, with the consequence that they build up in the fat of any animal which eats sublethal quantities of them. If a predator or scavenger then eats an animal with a store of insecticide in its fat, it may consume a considerable amount of the pesticide. If it receives enough in this way it will die. Secondary poisoning is the name given to this type of effect.

It is the combination of persistence and fat solubility in the pesticide or its metabolite which makes DDT, TDE, aldrin, dieldrin and heptachlor so unusual and such a potential threat to wild animals. These pesticides not only affect individual species but, by acting through food chains, can poison other species in the ecosystems of which they are part.

In 1960 the crucial characteristics of the persistent organochlorine insecticides just described were not perceived. Little was known about the insecticides other than that they were extremely effective in killing insects. Experience in the field had not yet raised any concern about DDT; it appeared not to threaten human beings and it had not been implicated with extensive kills of vertebrate wildlife. Yet, as early as 1945, Professor Wigglesworth (now Sir Vincent) had pointed out that DDT was unselective in its action and hence killed predatory insects which normally keep potential pest species in check. He emphasised that there were dangers in its use. However, no one seemed to have heeded this unpopular warning; the value of DDT was so great and so obvious that no one wanted to hear about any snags.

The recently introduced and more toxic insecticides (aldrin, dieldrin and heptachlor) were found to give good protection against Wheat Bulb Fly when they were applied as seed dressings. However, in 1959 and 1960 a number of farmers and game keepers reported that game and other birds were dying in the fields as the result of eating seed treated with these chemicals. Among the birds found dead were some predators. I related this fact with the Clear Lake story in which Western Grebes had died through feeding on fish carrying considerable amounts of the much less toxic insecticide TDE. It emphasised the need to study pesticides in food chains.

At the same time something strange was happening to Foxes in eastern England. Foxes were reported as dying in large numbers as early as 1958 and by 1960 the Master of the Fox Hounds Association had reported 1300 deaths: serious news for those people who enjoyed fox hunting. Many who did were influential and questions were soon raised in the House of Lords.

At first the deaths were put down to some unidentified disease. The Nature Conservancy was much concerned about the loss of an important predator and instructed Mr J. C. Taylor, their Warden Naturalist in charge of the Breckland National Nature Reserves, to investigate. He collaborated with the veterinary research worker Dr D. K. Blackmore of the Royal Veterinary College, London. Their field work showed that the fox deaths were almost certainly not due to disease, but to secondary poisoning by dieldrin and related insecticides; the foxes had eaten pigeons and other birds poisoned by eating corn dressed with the chemicals. These facts were made known to the Nature Conservancy in 1960 and there is little doubt that they engendered government support for further research on the effects of pesticides on wildlife, and hence for the support of Monks Wood Experimental Station. It was particularly fortunate that one of the first casualties of the persistent organochlorine insecticides was an animal whose welfare was so dear to the Establishment of the United Kingdom.

Surprisingly, the scientific reaction to the discovery of the cause of the Fox deaths was muted. Taylor and Blackmore published a preliminary note in the *Veterinary Record* in 1961, but their definitive paper did not appear until 1963, by which time interest had shifted to birds of prey, which (because of their relative rarity) posed a more urgent conservation problem. However, all should be grateful to Mr Taylor and Dr Blackmore for their pioneer studies on the Fox – the first animal in Britain to be a significant indicator species of a significant pesticide problem.

The argument about birds of prey was crucial in the long debate about the hazards of the persistent organochlorine insecticides and I shall discuss it in some detail. By 1961 there was growing evidence that some species of birds of prey were declining in Europe and North America. In that year my colleague Derek Ratcliffe organised the Peregrine Enquiry on behalf of the British Trust for Ornithology in order to discover whether Peregrines were having a serious effect on racing pigeons, as claimed by pigeon fanciers. The survey showed that there had been a considerable decline in the number of Peregrine occupied breeding sites and in breeding success. When an addled egg was found to contain 200 micrograms of persistent organochlorine insecticides and their metabolites (DDE, dieldrin, heptachlor, heptachlor epoxide and γBHC) Derek Ratcliffe and I independently came to the conclusion that the declining number of birds of prey was probably due to persistent organochlorine insecticides. The hypothesis appeared to be both reasonable and important so we published a note in *Bird Study* about it. The suggestion that bird populations might be affected by sublethal effects as well as lethal ones added a new dimension to the problem and confirmed our opinion that the effects of the persistent organochlorine insecticides were very complicated and especially worthy of study, and could not be understood without doing much research in the laboratory and the field. All we knew at the time was that these pesticides frequently affected individual mammals and birds as well as insects, and that they appeared to be affecting populations of some predators. We did not know which organochlorine insecticides were having the most significant effects nor how they affected vertebrates toxicologically. We did not know the relative importance of lethal and sublethal effects. We had no idea how many species of birds and mammals were seriously at risk. It was difficult to know where to begin when the research field was so huge. There was so little information.

The first requirement was to discover the scale of the problem. The only way to do that was to measure the amounts of pesticide in the bodies of as many animals as possible from as many parts of Britain as possible. This would tell us which species contained most and – discounting differences in response between species – which were therefore at greatest risk. Accordingly, we sought specimens of anything from anywhere. It was much easier to get specimens than to get them analysed. Our first analyses were by the laborious (and not very accurate) method of paper chromatography. They were undertaken by the Section's chemist, Mr Colin Walker. The recent invention of gas/liquid chromatography made analyses much quicker and much more accurate, although formidable problems of identifying particular compounds had to be solved. The apparatus required was expensive and we had no suitable place to put it since Monks Wood Experimental Station had not yet been built, and we were making use of temporary accommodation in Cambridge and at St Ives. To start with we had to rely on the help of others, notably the Laboratory of the Government Chemist in London. However, they could only analyse a limited number of samples for us and costs were high. Nevertheless a good range of specimens was analysed. In an ideal world we would have collected large samples on a random basis. In practice this was not possible: we had not the manpower to do the collecting, and many of the species in which we were particularly interested were so rare that collecting specimens might have damaged their populations significantly. If pesticides were killing individuals extensively, then animals picked up dead were more likely to contain larger residues than specimens collected at random. Therefore, even though many specimens had clearly been killed by something other than pesticides, our total sample was biased. Despite the unavoidable inadequacy of our sampling methods, patterns of great significance emerged. These will be discussed later.

The next stage in our investigation was much more difficult. We had to determine the significance of the residues which we were finding. Ideally we should have carried out toxicological experiments on the species which we were studying in the field. Unfortunately these were not easy to keep in large numbers under laboratory conditions, and it was uncertain whether any of them could be induced to breed in captivity. Therefore we had to do what medical toxicologists have always done – choose a suitable laboratory animal and extrapolate results obtained on it to the species we were studying in the field. In the first place Don Jefferies, our vertebrate toxicologist, used the Bengalese Finch (Fig. 39). It could be kept in large numbers and, unlike the domestic fowl, was sufficiently like its wild relations to stand in for a wild bird. However, toxicological work on one species can never be used as absolute proof that the chemical concerned will affect another species in exactly the same way. At best such work provides the basis for reasonable opinion. That has been good enough when the other species has been man: it had to be good enough when it was the Peregrine Falcon.

If individuals of a species were picked up dead and were found to contain pesticides in amounts which indicated death by poisoning, and if the population of the species was known to be declining where the pesticide was used, then there was strong

Fig. 39. Bengalese Finches (*Lonchura striata*). Toxicological studies on this species at Monks Wood Experimental Station helped to elucidate the complicated sublethal effects of persistent organochlorine insecticides, and with residue surveys and population studies of wild birds led to the phasing out of these chemicals.

circumstantial evidence that the population decline was due to the pesticide; but there was no absolute proof, for many things can be correlated without one being the cause of the other. Much the best way of being certain is to carry out an experiment in the field. Unfortunately field experiments on populations of birds of prey in a small and overcrowded island are virtually impossible. The populations of the birds are too small and dispersed, and there was no possibility of getting farmers over large areas to refrain from using the suspect pesticides so that we could compare results on their land with those on land where the pesticides were used. Therefore conclusions about the effects of pesticides on bird of prey populations had to depend on circumstantial evidence alone.

The situation as regards invertebrates and smaller vertebrates was quite different. We could hope to do field experiments on them and on the ecosystems of which they were part. So I spent much of 1961 trying to find areas where we could do work of this kind. However, no place could give us adequate control of the land, and my requests to the Nature Conservancy to buy or lease a farm encountered political obstacles and had to be abandoned. This meant that we would have to depend for experimental land on National Nature Reserves, on the small amount of farmland which we held at Monks Wood, and on *ad hoc* arrangements which we might make with farmers who were willing to help us with our field experiments.

To conclude, our research on persistent organochlorine insecticides had to include surveys of pesticide residues in wildlife, toxicological studies to discover the significance of those residues, observations on population changes of selected species and, where possible, experimental studies on populations. Only thus could we determine whether pesticides caused significant population declines or had fundamental effects on ecosystems.

The residue analysis and toxicology we could hope to do ourselves, but we would have to rely on others to collect most of our specimens in the field, to measure population changes and to provide land for experiments. Methodological and financial constraints determined what we could undertake; we had to do what was best with what was available. This was the background for our work on the effects of persistent organochlorine insecticides. We would have encountered the same problems and constraints if we had been studying the effects of any other chemical group. They are endemic to this sort of work and, to this day, inhibit research programmes which should be done on the effects of pesticides.

In the 1950s studies on the side effects of pesticides were largely incidental and anecdotal, and the need for international cooperation between the scientists concerned was hardly recognised. However, in 1960 the problems of pesticide use were discussed at the 8th Technical meeting of the International Union for the Conservation of Nature and Natural Resources (IUCN) which was held in Warsaw; afterwards IUCN set up a committee on the ecological effects of chemical controls. Professor John George, who was working with the United States Department of the Interior, and who had carried out some of the first field studies on the effects of pesticides on wildlife, was appointed as chairman. By 1963 research was gathering momentum, notably in the USA, the United Kingdom, the Netherlands, Sweden, France and the USSR, but most of the research workers involved had never met each other. I thought the time was ripe for an international meeting and in my capacity as secretary of the IUCN committee I approached the North Atlantic Treaty Organization (NATO) for funds, as I knew they gave support to gatherings which furthered international cooperation in science. NATO quickly agreed to finance an 'Advanced Study Institute on Pesticides in the Environment and their Effects on Wildlife', to be held at Monks Wood in 1965 under my directorship. There were virtually no constraints: when I asked if I could invite Russians, the reply was: 'Of course, but don't ask more Russians than Americans'.

I was determined to base the meeting on informal discussion rather than on lectures. At least one American could scarcely believe his fare would be paid without his having to give a talk. Of course, we did have lectures in order to provide introductions to the discussions, but there were never more than four a day.

A total of 71 scientists from 11 nations came to the symposium which lasted 12 working days. The participants stayed at Monks Wood, a nearby motel and in Girton College, Cambridge. Years later I met the wife of a distinguished Dutch professor. 'I know all about you', she said 'your wife helped my husband climb into Girton'. This was indeed true, for at that time the College porter locked the gate long before midnight, and our discussions had lasted long after what was considered an appropriate bedtime.

The programme included visits to other parts of Cambridge, the Huntingdon Research Centre, a chemical manufacturer's laboratory (Fisons at the Chesterford Park Research Centre) and the Scolt Head National Nature Reserve where, as we shall see, the terns and their eggs were contaminated with residues of dieldrin and DDT. The tide at Scolt was a little higher than expected and we were only just able to wade across. One of the taller Americans gallantly carried a lady from France over the channel while she held her umbrella aloft. On the way back we dined at Ely and several participants took the opportunity of going to a music recital in the Cathedral clad in their very wet clothes. All survived a very memorable occasion.

On our last day we drew up a general statement for public use. It made pleas for monitoring both pesticide residues and populations of animals which might be affected, for standardising techniques of chemical analysis, for experimental research on the dynamics of pesticides, for long-term studies in the field and for better control. In all these subjects our main concern was with persistent pesticides.

By the end of the two weeks we had all got to know each other very well. Individual research workers felt less isolated, and future cooperation among those working on the environmental effects of pesticides was assured. The initiative sponsored by NATO was followed up by the Organization for Economic Cooperation and Development (OECD). This organisation promoted a series of meetings on the 'Unintended occurrence of pesticides in the environment'. The first was held at Jouy-en-Josas near Paris in 1966. Subsequently analytical procedures were checked by sending samples containing known but not divulged residues to different laboratories in different countries for analysis. The range of replies gave a good indication of the accuracy we could expect in any future monitoring schemes. National monitoring schemes were noted and the OECD group organised its own, using mussels to monitor pesticides in the sea, pike to monitor pesticides in fresh water and starlings in the terrestrial environment. While mussels and pike gave reliable results for local areas, starlings (which had been chosen for availability) were less useful since they often travelled large distances and fed in many places: they could only give a general indication of pesticide contamination. All this work drew attention to the widespread occurrence of persistent compounds, for governments as well as scientists, and led to pressure for their better control.

The Monks Wood Symposium was significant in being the first of its kind. It was followed by many others of varying nature and under many different auspices. Increasingly these meetings brought the makers and users of pesticides into contact with those who strove to make their use safer. These were exciting times for those of us who worked on pesticide effects at Monks Wood. We had dropped a small pebble into the pool but the ripples it caused were reaching the furthest shores.

15

*

The results of research

Research is never as tidy as it appears when read about in a scientific paper or article. Few programmes are planned in the logical and orderly sequence in which they are described. Ours was no exception. Constraints and opportunities determined the details of what we did and when we did it. We advanced irregularly on a wide front – residue studies were started in 1961 and still continue at Monks Wood. Field experiments on the effects of a specific pesticide on an ecosystem were started in the same year. Toxicological studies had to wait until specially built laboratories were available at Monks Wood in 1963. Studies on changes in bird populations were achieved through the cooperation of the British Trust for Ornithology (BTO) and its network of hard-working volunteers, but the requirements of the BTO determined the timing of the surveys. All this should be remembered when the following, deceptively tidy, account of our research is read.

The Nature Conservancy was and is concerned about all wildlife – plants and animals, vertebrates and invertebrates, rare species and common species – but it has to give priority to the threatened species if it is to maintain genetic diversity. Since large animals tend to have much smaller total populations than small ones and tend also to breed more slowly, on average they are more at risk; so, while studying the effects of pesticides on many types of creature, we tended to give priority to work on large animals (notably birds).

We obtained specimens for residue analysis from many sources. Once people knew about our studies we were sent a variety of corpses from landowners and naturalists who were curious to know what had killed the mammal or bird which they had found dead. Strange and sometimes unsavoury parcels came through the post. One person sent a fox on the day before a postal strike: it cannot have been popular at the sorting office, and when it did eventually reach us its value as a specimen was considerably reduced. Analytical constraints forced us to be very selective; we concentrated on predators. We put out requests for predatory birds through the good offices of the RSPB. There was no question of being able to shoot specimens of Peregrine and the like and so it was imperative that we made use of any dead bird that was found. It was an exciting period: we awaited every report from the analysts with intense interest. Almost every specimen suggested some new possibility, some new hypothesis. We expected that birds and mammals living in the agricultural lowlands of England would often contain residues of persistent organochlorine insecticides or their metabolites, and they did. Initially we did not expect to find detectable residues (i.e. more than 0.1 parts per million) in every land bird and mammal which we looked at. But not only did we find residues in species like Wood Pigeons (which we knew fed on contaminated grain or plants) but also in birds which ate insects, and in predatory mammals and birds of prey (including owls) which fed on mammals and birds.

The pesticides were clearly also getting into the lowland rivers and ponds, because we found them in all the fresh water species which we studied. The Zebra Mussel (*Dreissena polymorpha*), is a common species in the River Ouse and other fenland rivers. It was originally introduced from the Caspian region about 150 years ago, but thrives in Britain. Specimens from the Ouse contained residues of DDE, the main metabolite of DDT, however, not in sufficient quantities for the Zebra Mussel to be the useful monitoring organism which I had hoped it would be. On the other hand, the first pike which we analysed contained a good deal of pesticide. This rather worried one of my colleagues who had made an excellent meal on those parts of the fish not required for chemical analysis, *before* he had received the results from the chemist. He suffered no ill effects.

We searched further afield for pesticide-free animals – notably in upland areas where little or no pesticides were used. The

Ptarmigan is an arctic-alpine bird which feeds on bilberry and other plants high up above the tree line. If any land bird were to provide a negative result this was it. Even so, a Ptarmigan egg from the Scottish highlands contained DDE. We continued our search and eventually I succeeded in finding a specimen with no detectable residue. It was a trout which I had tickled in a small burn in the Speyside mountains; the chemist could find no trace of persistent organochlorine insecticide residue in it. Admittedly it provided a very small sample, and I have to confess that it could have been larger (I caught two others, but I took them back to my colleague and host, Frank Green, and we had them for breakfast next day at Anancaun).

In 1963 Max Nicholson sent me to the USA to study the situation there. In the course of a visit to the laboratory of the Department of the Interior at Denver in Colorado I heard that DDT had been found in a sample of cod liver oil. This seemed an extraordinarily interesting finding but little significance seemed to have been attached to it in America. Later Dr Michael Way, the first botanist to join the Toxic Chemicals and Wildlife Section, told me that he was going to visit the North Norfolk Coast and he offered to pick up specimens for me for analysis. 'Even sea birds' eggs?' he asked. Remembering the cod liver oil I replied enthusiastically 'Yes'. Sure enough, considerable residues of several persistent organochlorine insecticides were found in the addled sea birds' eggs which Michael Way brought back (Fig. 40). At that time no one had detected DDT in sea water but our results from sea bird eggs showed that it must be there. Since the terns and other sea birds of Norfolk feed in waters receiving agricultural runoff from the fenland rivers it was not really surprising. We then analysed Guillemot and Shag eggs, which Derek Ratcliffe had brought us from St Bees Head in Cumberland and St Abbs Head in Berwickshire. These too contained substantial residues of pesticides; it was clear that all the British seas must be contaminated with persistent organochlorine insecticides. In fact the amounts of pesticide found were so large that it was possible to use sea bird eggs for monitoring changes in the amount of persistent organochlorine insecticides in the sea. We started to use them for this purpose in 1963.

Our preliminary studies were revealing such interesting results that we decided to make them widely known, so Colin

Walker and I published an account of them in a paper to *Nature* in 1964. Later that year at Swansea I gave a talk at the British Ecological Society's symposium on ecology and the industrial society. I emphasised the significance of finding persistent organochlorine insecticides in sea birds. In the evening following my talk a journalist from the *Sunday Telegraph,* which had been recently launched, rang me up asking for details: the widespread occurrence of persistent organochlorine insecticides in the environment had become a matter of general interest. Our reconnaissance survey of pesticide residues in wildlife had shown that virtually all living things in Britain were now in contact with these new man-made toxic substances – this was indeed news. We had demonstrated that pesticides were a new ecological factor which would have to be taken into account in all future studies of wildlife in Britain. The implications for conservation were obvious and the economic ones were considerable, because pesticides were present in the estuaries and shallow seas, and these were the breeding places of many species of fish on which the fishing industry depended.

Fig. 40. Nest of Sandwich Tern (*Sterna sandvicensis*), Scolt Head National Nature Reserve. The discovery of residues of persistent organochlorine insecticides in the eggs of sea birds demonstrated that British waters were contaminated by these chemicals. Sea bird eggs were subsequently used to monitor changes in contamination by pesticides and hence the efficacy of measures to restrict their use.

The amounts of pesticide found varied from specimen to specimen. For example, some pigeons contained little, some had so much that it was probable that the birds had died of pesticide poisoning. Yet, despite the inevitable inadequacies of our sampling techniques, a clear pattern was emerging from our studies of pesticide residues in birds. Colin Walker and I drew attention to it in our paper to *Nature* in 1964: in the aquatic environment, species which fed on fish (Herons and Great Crested Grebes) contained much larger residues than Moorhens which fed mainly on plants. Similarly, in the terrestrial environment, the flesh-eating predators contained larger residues on average than thrushes which fed on plants and invertebrates and Wood Pigeons which were entirely herbivorous. There was also an indication that the predatory birds which ate other birds contained more than those which fed on mammals. These findings were consistent with the fact that while persistent organochlorine insecticides were stored in the fat of all animals, they were more rapidly broken down by mammals than birds. At the time we wrote our paper, the samples were very small; however, analyses in later years confirmed the patterns which we had found to a large extent.

Many people, including ourselves, had concluded from Hunt's and Bischoff's study at Clear Lake in California that the amount of pesticide present in a species depended upon its position in the food chain. In other words, persistent pesticides became concentrated in food chains with the result that species at the ends of food chains contained most and so were at greatest risk. Indeed at Clear Lake it was the Western Grebes, which had eaten the fish which had eaten the plankton, which had died. The results of our survey seemed to confirm this hypothesis: the predatory pike contained more residues than the other fish, Herons and Grebes more than most of the fish which they ate, falcons and hawks more than the species they ate. It was obvious that we should concentrate on those species with large residues at the ends of food chains and we did precisely that. We paid particular attention to Peregrine Falcons, Sparrow Hawks, Golden Eagles, Barn Owls, Kestrels, Herons and Great Crested Grebes.

It was a most fortunate accident that the British Trust for Ornithology carried out its survey on the Peregrine Falcon at the time when that species was at its lowest ebb. Derek Ratcliffe soon realised that the effect of the pigeons on the Peregrines was much greater than the effect of the Peregrines on the pigeons, for many of the pigeons contained residues of persistent organo-chlorine insecticides. It was Derek Ratcliffe's meticulous study of the Peregrine which provided the best evidence of the effects of these pesticides on one species. Meanwhile, Mr Ian Prestt (a colleague of mine from Regional work days and later the Director of the RSPB) organised surveys on the other predators: a careful opinion poll of BTO members on the status of Sparrow Hawks, Kestrels and Barn Owls, and a repeat of the 1931 study of the Great Crested Grebe population by P. A. D. Hollom, also through the good offices of the BTO. Golden Eagles and their breeding success were studied by Dr James Lockie, then with the Nature Conservancy in Scotland and later a lecturer at the Department of Forestry in the University of Edinburgh. Ian Prestt, Mr Tony Bell and Mr Peter Milstein studied some heron-ries in eastern England intensively, while the regular surveys of the total English and Welsh population of Herons by the BTO continued to provide the crucial data on their fluctuations. The study of Sparrow Hawks was backed up by a study on their breeding success in selected localities in England and Wales by Ian Prestt and Tony Bell, and later the studies of Dr Ian Newton on Scottish populations of this species added much more infor-mation about the effects of pesticides upon it.

All these studies have been fully described in the scientific literature and in Derek Ratcliffe's admirable book on the Pere-grine. My purpose here, as it was in the 1960s, is to relate the studies to each other and to draw conclusions which are relevant to conservation and the control of pesticides. All the species which we studied shared the common characteristic of being at the end of the respective food chain and of containing more persistent organochlorine insecticide residue in their bodies and/or their eggs than the average. However, the studies on their populations showed that there were great differences between them and hence in the effects pesticides were having on them.

In the period between 1931 and 1965 the Great Crested Grebe population had increased by 137%. If any individuals had been killed by pesticides this was greatly outweighed by some other factor which was enabling the species to increase. This factor was almost certainly the rapid growth of the gravel extraction industry whose byproduct was numerous water-filled pits. These provided a great increase of particularly suitable new habitats for Grebes during the period when the use of persistent organochlorine insecticides was expanded.

A number of Herons analysed contained residue levels of dieldrin which strongly suggested that the birds had been killed by this insecticide. Further, Ian Prestt and his colleagues found that many of the eggs in the heronries which they were studying in the Fens and Lincolnshire failed to hatch, and that young birds were being killed by their parents throwing them out of the nest. (Later Don Jefferies demonstrated that dieldrin could produce a similar effect in Bengalese Finches in the laboratory.) Yet, despite these casualties, the BTO survey showed that no unusual decline in the English and Welsh population of the species had occurred during the period the suspect insecticides had been used. The very cold spell at the beginning of 1963 had had a dramatic effect on the Heron population (as on many other species) but, within 7 years, it had recovered to its normal level (1928–77 mean) of approximately 4000 pairs. The most that pesticides had done was to cause a slight delay in recovery after the cold winter. Clearly the Heron was able to make good any losses that had been caused by pesticides.

The Barn Owl had undoubtedly declined in recent years, but the extent to which this had resulted from poisoning by persistent organochlorine insecticides could not be determined. Barn Owls feed principally on small mammals but, as we have seen, laboratory studies had shown that these animals were able to metabolise the pesticides more rapidly than could birds, hence the owls were less likely to feed on animals with large residues, and indeed the residues found in them rarely suggested that they had died from poisoning by these particular chemicals. It was more likely that the decline of the Barn Own was due to overall changes in farming practice (including the use of pesticides) which reduced their food supply and to a decline of suitable nesting sites. It is possible that some were the victims of secondary poisoning by rodenticides. This subject needs more study today as more toxic rodenticides are used to replace warfarin, to which many rats are now resistant.

The Golden Eagle's population of *c.* 200 pairs had not declined during the period under review, but its breeding success had become much reduced in those parts of Scotland where the eagles were feeding on dead sheep. Dieldrin was found in Eagle corpses and eggs and it was presumed that the eagles obtained the insecticide from sheep which had been dipped in dieldrin in order to protect them from fly. Derek Ratcliffe also showed that the egg shells of eagles had become thinner like those of Peregrines (see below). Thus it appeared that pesticides were not yet affecting the breeding population of these long-lived birds but that, if breeding success continued to decline and there were inadequate replacements of adults, the species was threatened with ultimate extinction.

Ian Prestt's BTO study of the Kestrel showed that it had declined markedly in those parts of England where arable farming predominated. In the lowlands only those Kestrels living in the towns and suburbs seemed unaffected. Similarly, the Sparrow Hawk disappeared from huge areas of lowland Britain. This species, which used to occur in nearly every wood in the country despite the persecution of game keepers, had become extinct over hundreds of square miles. On the fringes of its much reduced range it had very poor breeding success and its egg shells had also become thinner. Oddly enough it was difficult to get formal evidence that the Sparrow Hawk had declined. Everyone noted its disappearance but very few detailed records had been made of its distribution and numbers when it had been abundant: it was so common that ornithologists did not bother to record it in their reports. Continuation of the BTO's annual Common Bird and Woodland Bird Censuses will ensure that this will not happen again: in the future we shall have records of common species as well as rare ones.

By far the best-documented decline in both population and breeding success was that of the Peregrine. From a pre-war population of about 874 pairs the species had declined to 44% of that number by 1964. It became extinct in southern England and most of Wales. The only normal population was that in the Central Highlands of Scotland. Breeding success in the surviving pairs declined. Derek Ratcliffe' classic study on this species showed that, as soon as the species came into contact with DDT, its egg shells became significantly thinner on average than they had been in the pre-war period.

The picture which emerged from the studies of these seven species of pesticide-contaminated predators was complicated. The effects of the pesticides on populations appeared to range from none in the Great Crested Grebe and Heron to severe effects in the Kestrel and Sparrow Hawk and to very severe effects in the Peregrine, and they posed a dangerous threat to the Golden Eagle. Emphasis must be put on the word 'appear'. To those of us who knew something about the natural history of these species and had observed them in more normal days, the unusualness of the population declines and their close correlation in space and time with pesticide use made it virtually certain that the persistent organochlorine insecticides, which we were finding in the bodies and eggs of these species, were the cause of the declines. Yet we had done no experiment which excluded other plausible causes. As we ourselves had frequently pointed out, the 1960s were a period of rapid agricultural change. Theoretically, the declines could be attributed to shortage of food, to disease or to increased human disturbance – and they were thus attributed because many people did not want to believe that the otherwise very valuable insecticides were damaging, and were very anxious to put forward other reasons for the decline of the affected species. In any case, as scientists, we had to test our own conclusions by examining very carefully the alternative hypotheses for the declines. Only food shortage could be dismissed easily since all the species concerned ate a wide range of foods, and wild small birds and mammals were still abundant on most land; with the possible exception of the Barn Owl in places where agriculture was very intensive, it was inconceivable that predators could have been exterminated over large areas of countryside by lack of food.

With increasing prosperity there was an increase in game preservation, hill walking and rock climbing. We knew of cases where these activities had been harmful to individual birds but, again, it was inconceivable that their increase was on a scale to exterminate species over hundreds of square miles.

To prove that disease was not the cause of the declines was more difficult, as Taylor and Blackmore had found with the fox. We received much help from the veterinary laboratory of the Ministry of Agriculture at Lasswade who attempted to determine the cause of death of the specimens obtained for chemical analysis which we sent them. There had been outbreaks of

Newcastle disease in wild birds, but their timing did not fit the declines of the birds which we were studying. Some of the birds which were examined had died of disease, but there was no evidence that there had been a pandemic on the scale of myxomatosis, which alone could have accounted for the disappearance of the birds of prey on the scale observed. To us the circumstantial evidence for stating that pesticides were the cause of the population declines was overwhelming.

There is no doubt that if we could have demonstrated the exact physiological mechanism by which the persistent organochlorine insecticides killed birds that would have gone a long way to convince those who doubted our interpretation of events in the field. Great efforts were made by many research workers to determine how these insecticides actually work, not least by Don Jefferies and Arnold Cooke in our Section. Much knowledge was gained but, to this day, there is much debate on how DDT actually kills insects, and how its metabolite DDE causes the thinning of egg shells in birds. The problem is that DDT has numerous effects and it is difficult to determine which causes the primary lesion and so can be identified as the cause of death. Even so, enough is known today for us to give a reasonable opinion about the relative importance of lethal and sublethal effects of pesticides. In Britain there is little doubt that the population declines of Peregrine, Sparrow Hawk and Kestrel were primarily due to acute poisoning by the closely related and very toxic insecticides aldrin and dieldrin. The recent work of Ian Newton and his colleagues leaves little doubt that DDE, the metabolite of DDT, is the chemical mainly responsible for egg shell thinning and that dieldrin at least can cause behavioural anomalies which can also affect reproductive success. In Britain the sublethal effects of these pesticides were not the primary cause of the declines in population but, by reducing breeding success, they have almost certainly delayed the recovery of the species concerned.

The final 'proof' that the organochlorine insecticides were the cause of the declines of the birds of prey eventually came when it was no longer politically necessary – when the pesticides concerned had been withdrawn from extensive use. Clearly, if the pesticides had been the cause of the declines, the birds would recover when the insecticides were phased out. First we should see a fall in the amount of pesticides in the bodies of the birds and their eggs, then a recovery in breeding success and, finally, an increase in population. We predicted that all these trends would be seen in all the affected species when the pesticides were withdrawn from use, and they were. The egg shells of the Golden Eagle returned to normal and so did their breeding success. The Sparrow Hawk and Peregrine slowly but surely returned to many of their old haunts in England and Wales. The process of recovery is not yet complete – Sparrow Hawks are still rare in East Anglia and Peregrines are still absent from some of their old breeding places on the coast of the English Channel. There is every reason to suppose that, when the last vestige of the persistent organochlorine insecticides has disappeared, they will have returned to all parts of their former range in Britain.

Research on birds of prey was important because of the intrinsic interest and conservation value of the species concerned, and because it showed how a few persistent insecticides could affect a vertebrate species over hundreds of square miles. Indeed the effects of these chemicals on the Peregrine were worldwide. In the eastern United States it was totally exterminated – not as in Britain by dieldrin, but by DDT which was used on a far greater scale in America than in Britain. The Peregrine disappeared from much of Europe. Although the use of persistent organochlorine insecticides was banned in Finland, its breeding population of Peregrines declined from at least 1000 pairs to less than 20 pairs by 1977. Finland is too cold in winter for Peregrines and so they migrate south to countries where these chemicals are still used. This explains why the Finnish population has not recovered like the population in Britain which does not need to migrate.

Birds of prey can never be very abundant since they are at the ends of food chains and require many prey animals to support them, so the total effect of a bird of prey species on the ecosystem in which it occurs can never be very important. Insects on the other hand are far more numerous and more likely to be important ecologically. Therefore it seemed very desirable to discover what effects the persistent organochlorine insecticides might be having on them. Because insects are so numerous it is difficult to count enough of them to make valid estimates of their numbers in large areas and hence to have any real knowledge about their fluctuations. However, some insects, notably dragonflies and butterflies, are large and conspicuous and can be counted in the field. As a result we can have a rough idea about their recent population changes. It seemed possible that we could determine the effects of pesticides on them at least.

In the 1960s many naturalists noted a decline in butterflies. 'Where have the butterflies gone?' became a journalistic headline. Knowing that the use of insecticides of all sorts was increasing and hearing about the possible effects of the persistent organochlorine insecticides on predatory birds, many people believed that butterfly populations were declining because of the insecticides. Again the Nature Conservancy had to investigate the truth or otherwise of the matter. The Section's toxicologist who dealt with invertebrates, Dr Frank Moriarty, carried out some laboratory experiments on the Small Tortoiseshell Butterfly and discovered what dose levels of pesticide killed them or had effects on their reproduction. From this work he concluded that butterflies which were directly sprayed would be affected but that it was most unlikely that butterflies living outside the crop areas would be seriously affected. In other words, something else was causing the general decline of the butterflies.

For the grassland species – the Browns, the Blues and the Skippers – the cause was obvious. Thousands of acres of old pasture which contained the food plants of these species were being ploughed up and replaced by grass leys of Italian Ryegrass on which no butterfly caterpillars can feed. Most of the fields which were not ploughed up were treated with selective herbicides with the same general effect. It was less obvious why the fritillary butterflies, whose caterpillars live on violets in woodland, had also disappeared from most of eastern England. This problem has not been resolved but it is probably due to the neglect of broad-leaved woodlands which for years has been general over much of the country. Under the old coppice and standards rotation, the herb layer of plants (including violets) was allowed to flourish at regular intervals whenever the coppice was cut. The violets became much rarer or disappeared when the coppice ceased to be cut, and so the fritillaries lost their food plant. Also the widespread use of insecticides on farmland between woods may have made it increasingly difficult for fritillaries to disperse safely from one wood to another. Our general conclusion was that the disappearance of butterflies was not due primarily to insecticides but to changes in grass husbandry and woodland management.

Pesticides of all kinds were clearly affecting the weed flora and the invertebrate fauna of croplands. We felt that it was highly desirable to study the total effects of pesticides on the flora and fauna of arable land but, as mentioned above, we were unable to obtain land suitable for the purpose. However, we were able to do two experiments on the effects of pesticides on ecosystems. The first was a theoretical one: it threw light on the way persistent organochlorine insecticides might affect an ecosystem.

I had learnt from Charles Elton that the woodlouse-like Pill Millipede *Glomeris* (Fig. 41) was instrumental in breaking up the dead leaves of Tor Grass (*Brachypodium pinnatum*) so that microorganisms could break it down further and make nutrients available in the soil. In other words, it appeared that the whole Tor Grass system depended on the action of one species of millipede. I thought it would be of great theoretical interest if we could test this hypothesis by killing off the millipede with an insecticide and observe what happened.

Brian Davis and I carried out the experiment in a detached bit of the Castor Hanglands National Nature Reserve and Colin Walker assisted us with chemical analyses. We found a dose rate of dieldrin which was successful in killing off the *Glomeris*

Fig. 41. Pill Millipede (*Glomeris marginata*). A field experiment on the effects of dieldrin on a population of this animal showed that pesticides could be used to understand the functional role of species in an ecosystem (enlarged).

without affecting other species. As predicted, the litter of dead grass did build up to some extent but then other millipedes seemed to increase and to fill the role of *Glomeris* and so the Tor Grass system was little affected by the insecticide. This study demonstrated the buffering ability of an ecosystem and suggests one reason why arable ecosystems have been less altered by continuous insecticide use than might have been expected.

The other study of the effect of a persistent insecticide on animal relationships was that undertaken by Dr Jack Dempster on the effects of DDT when applied to Brussels sprouts to control the cabbage white (*Pieris rapae*). He found that when the crop was first sprayed, DDT gave some control of the pest but, by a year later when the crop had been sprayed three times, there were more than three times as many caterpillars on the sprayed crop than on the unsprayed control. A careful study of the animals which preyed on the caterpillars showed why this was. When the crop was not sprayed, 90% of the eggs, larvae and pupae of the cabbage whites were eaten by the ground beetle *Harpalus rufipes* and the harvestman *Phalangium opilio* (Fig. 42). When DDT was applied, it killed enough of these predators to

Fig. 42. Ground beetle (*Harpalus rufipes*) and harvestman (*Phalangium opilio*). Field experiments showed that when DDT was applied to control cabbage white butterflies on brassica crops the populations of these predaceous species were reduced and this led ultimately to an increase in the pest species.

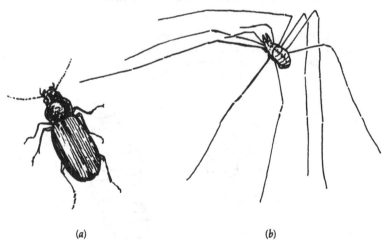

(a)   (b)

reduce greatly the control they were exerting on the cabbage whites; when new butterflies came in and laid eggs on the Brussels sprouts after each spraying, the caterpillars which hatched from their eggs had much fewer predators to contend with than those on the unsprayed control plants. Jack Dempster also showed that even when DDT did not kill the ground beetles, it reduced their rate of feeding, thus reducing the efficiency of the beetles that survived in controlling the cabbage whites.

Throughout the period, when all these studies were being made, the simple model of accumulation of pesticides in food chains, provided by Hunt and Bischoff in their pioneer work at Clear Lake, had influenced our thinking; as more research was done on the effects of persistent organochlorine insecticides by laboratories throughout the world it became clear that it was inadequate to account for what happened in many situations, not least in Clear Lake itself. For example, in a laboratory study on earthworms and Song Thrushes at Monks Wood, Brian Davis and Don Jefferies showed that while persistent organochlorine insecticides did indeed 'flow through' food chains, the amounts at each trophic level did not necessarily increase. They found that the earthworms might accumulate more pesticide than the thrushes at the end of the food chain yet the lesser amounts in the thrushes were enough to kill them.

As more studies were made on how aquatic animals obtained pesticide residues from the environment it became clear that molluscs and fish picked up most of their pesticides through the gills rather than through their food. Therefore it was almost certain that the fish, which had provided the TDE which killed the Western Grebes on Clear Lake, had absorbed most of the chemical in this way rather than through their food.

Our studies at Monks Wood and the increasing number which were being made elsewhere emphasised how complicated were the effects of the persistent organochlorine insecticides on wildlife. There were and still are great gaps in our knowledge, but enough had been done to indicate that there were real hazards in using these chemicals and so their use should be restricted and eventually withdrawn altogether.

16

*

Peregrines and people

The research at Monks Wood on the effects of persistent organochlorine insecticides produced results which were biologically interesting, but that was incidental: the work was designed specifically for practical conservation reasons – to reduce pesticide hazards to wildlife. Those in the Nature Conservancy had a dual role: we had first to do the research, and then we had to see that our findings were applied. In this we differed from most of those working on strictly agricultural subjects; for them, one organisation, the Agricultural Research Council, did most of the research, while another, the Ministry of Agriculture, Fisheries and Food (MAFF), put it into practice through its advisory service. However, parts of the Ministry operated more on our lines: in particular the Pest Infestation Control Laboratory, who did research on pest problems in order that the Regional Pest Officers could give advice about solving them.

By 1961 casualties due to birds and mammals eating corn which had been dressed with aldrin, dieldrin and heptachlor were numerous, and had been recorded not only by game keepers, naturalists and ourselves but also by MAFF officials. No one knew what the effects were on populations, but enough individuals were being killed to cause serious alarm among the interested groups and the general public. Many more casualties were found at spring than at autumn sowing. Presumably there were more alternative foods in autumn so birds were less attracted to corn lying on the soil's surface; also they had more fat in their bodies. This was important, because, while the pesticides remained in the fat they had no physiological effects, but when fat was mobilised at the end of the winter, the pesticide residues were discharged into the blood stream and affected the central nervous system. By 1961 the damage to wildlife was so obvious that the Advisory Committee on Poisonous Substances used in Agriculture and Food Storage (later called the Advisory Committee on Pesticides) quickly accepted the recommendations of its wildlife panel that aldrin, dieldrin and heptachlor should no longer be used as seed dressings for spring sown corn. As a result of this voluntary ban on one specialised, but very important pesticide use, the number of wildlife casualties declined markedly.

While welcoming this restriction on one use of the more toxic persistent organochlorine insecticides, we did not feel that it went far enough. The discoveries which we had made in 1962 and 1963 gave us much concern about the total use of all such insecticides. Our work on birds of prey, notably Derek Ratcliffe's research on the thinning of egg shells, suggested that sublethal effects of these compounds might be much more important than had been realised. Secondly, our studies on pesticide residues were showing extensive pollution of all the environments at which we looked, including the sea. If virtually all wildlife was in contact with persistent organochlorine insecticides and these had significant sublethal effects as well as lethal ones, the danger to wildlife might be very considerable, if not catastrophic. Moreover, surveys by the agricultural departments showed that the use of persistent organochlorine insecticides was increasing and so the dangers would increase. By 1963 all of us at Monks Wood who were working on the effects of these pesticides on wildlife had no doubts that their use should be greatly restricted and eventually banned altogether. How was this to be achieved?

Industry had agreed to abide by the decisions of the Advisory Committee on Pesticides under the Pesticide Safety Precautions Scheme. As mentioned above, the Nature Conservancy was formally represented on this committee by Bob Boote, while I had been appointed to its scientific sub-committee and wildlife panel. Clearly the proper channel for our representations must be through these committees. However, we soon met considerable opposition to our views on the committees and it was becoming clear that we had to appeal to a wider audience as well. We wanted open, well-informed national debate, because increasingly we saw the persistent organochlorine insecticides as a special example of a much wider phenomenon, which urgently required scientific discussion and governmental action. To us the persistent organochlorine insecticides seemed to provide a warning of the dangers of persistent pollutants in general. This view was soon to be vindicated when it became possible to identify residues of a class of industrial pollutants – the polychlorinated biphenyls (PCBs), and we found that they, like the insecticides, were widely distributed on land, in fresh water and in the sea (see Chapter 17).

Early in the 1960s, the Swedes demonstrated that mercury compounds used to control moulds on timber, and fungal pests on cereals could cause serious casualties to wildlife in water and on land, respectively. Therefore we were also concerned at Monks Wood about pollution by heavy metals like mercury, which (unlike the pesticides and PCBs) could never break down into harmless constituents (see Chapter 18).

We knew very well that we must warn everyone about the dangers of all persistent toxic chemicals, because everyone was involved; even if most people were not working in agriculture, fisheries or conservation, everyone ate food which could be contaminated. Those of us who earned our living by conserving nature won considerable satisfaction from the fact that our work was helping to conserve man as well as wildlife, especially at a time when most people felt that conservation was a somewhat trivial activity. I felt particular satisfaction about this as my father and grandfather had been medical men and there were times when I thought I should have followed in their footsteps.

The proper way for scientists to propagate their ideas is to write up their work in scientific journals, reviews and books. We did this, and I believe that our wish to warn people of the actual and potential dangers of persistent chemicals helped to keep us up to the mark in publishing our findings. It is sadly interesting that, when Monks Wood lost its role as a promoter of conservation, its publication rate declined significantly.

Sometimes the press got hold of one of our scientific papers and produced an article about it. The wish to tell a good story frequently meant that the original work was distorted or exaggerated. This was distressing since we naturally wanted to keep our reputation of objectivity. As a result of such distortions by the media, we developed an ambivalent attitude to the press and broadcasting and television organisations. We welcomed the fact that they brought our work and its message to a wider audience, but we were embarrassed by the inaccuracies, especially when fellow scientists believed what they read in the papers rather than what we had said ourselves – a thing they would never do when their own work was reported in the press!

In later years people often said to me, 'I suppose the pesticide work at Monks Wood was triggered off by Rachel Carson's book *Silent Spring?*' Actually it was not; we were rather proud of the fact that we had started our work on the persistent organochlorine insecticides some time before this famous book was published. I remember my first contact with it well. I was just about to investigate the harmful effects of an oil dispersant on the shore animals in Milford Haven, where there had been a serious oil spillage. I knew the area well, having used it to teach students marine biology when I was a lecturer at Bristol University. Just before I left for Pembrokeshire a large official parcel arrived with 'urgent' written all over it. I was rather surprised to find that it contained copies of the *New Yorker*. Somehow I did not expect my boss, Max Nicholson, to send me humorous reading matter to alleviate field work in Milford Haven. The covering note explained all. Lord Hailsham, who answered for us in Parliament, had heard about the great stir which Rachel Carson's articles were causing in the USA – publication in the *New Yorker* had certainly not lessened their effect! Lord Hailsham wanted our views on them quickly, and so Max Nicholson had sent the *New Yorkers* to me with instructions to assess the relevance of the articles to the situation in Britain. Plans were modified, much midnight oil burnt, and the deadline fulfilled.

Silent Spring was soon available in book form in the United Kingdom. It had a considerable impact and, although it did not alter our research programme, it did form many people's views about pesticides and their views about the work we were starting at Monks Wood. We received a considerable boost from the association, which almost certainly helped us to get more resources for our work. By the time I visited the USA in 1963, Rachel Carson was a very sick woman. My American colleague on the pesticide committee of the IUCN Professor John George, offered to introduce me to her, but she was too ill to be visited so, sadly, I was unable to thank her personally for the help she had given us.

The 1960s are now sufficiently long ago for it to be possible to look at them in historical perspective. They provided the environment in which we did our research and promoted its results. The decade immediately before the oil crisis of 1973 was a strange one and I suspect it will seem increasingly aberrant as the years go by. Harold Macmillan's assertion that 'we never had it so good' might have been true in economic terms, but the gradual realisation that prosperity was being achieved at the cost of the environment weighed heavily on many people. There was a growing feeling that the dream could end in nightmare. The work which we and others were doing on pollution reinforced this feeling of doom. However, in general, the 1960s were almost Victorian in their air of optimism. Youth was liberated and having its fling and the liberation of women seemed to promise much more than sexual freedom. So, while prosperity was tempered with anxiety about the future, there was confidence (at least among the young) that the environmental problems could and would be solved as the facts became known. At Monks Wood we had no doubt that it was our duty to ensure that the results of our research on pesticides were applied. It was obvious that the persistent organochlorine insecticides should be phased out. Naïvely we assumed that it would be easier to get a ban on these chemicals than it had been to show that they were dangerous. We prepared for Conservation Year, the Countryside in 1970 and the Stockholm Conference of 1972 with confidence that we could do something effective about the new environmental problems.

Splendid generalisations were made: not seeing the wood for the trees is a common fault of scientists, since their work must concentrate on detail, but the 1960s were good for seeing the wood. Too good in fact: we needed the support of that practical visionary, William Blake, who said

He who would do good to another must do it in Minute Particulars, General Good is the plea of the scoundrel, hypocrite and flatterer; For Art and Science cannot exist, but in minutely organised Particulars.

It was no good accepting the general idea that persistent pollutants were dangerous if we did not reduce environmental contamination by particular substances. We had to start applying our research with regard to the chemicals we knew most about – the persistent organochlorine insecticides.

Like most scientists who were new to politics I thought that changes in opinion and action were mainly produced by obtaining facts and arguing logically from them. I was soon disillusioned. I found that there were immense obstacles to implementing the general restrictions we wanted. Facts and logic were only the first phase; success would depend on overcoming economic and political obstacles.

The obstacles to a ban were formidable. All the persistent organochlorine insecticides were extremely effective agents of insect control. DDT had saved millions of lives by controlling the vectors of a number of important diseases, and all of these pesticides – in the short term at least – had saved many other lives indirectly by controlling insect pests of crops. In the case of DDT, these miracles had been achieved with the loss of few if any human lives since its toxicity for man was so low. The more toxic aldrin and dieldrin had caused a few fatalities, when seed corn dressed with them had been eaten by mistake, but the losses were small compared to the benefits obtained. Few people wanted to hear about the disadvantages of the persistent organochlorine insecticides when the advantages were so great, least of all those who earned their living by making, selling or using them.

The production of pesticides grew almost exponentially in the 1960s: they became big business. The profits for manufacturers from the sale of persistent organochlorine insecticides in the United Kingdom were considerable, but they were insignificant compared to the sums received from sales overseas. The reduction of sales of these chemicals in the British market, though commercially undesirable, was itself bearable, but it would be very serious for the chemical industry if the much larger overseas market were affected. Restrictions in Britain were bound to have effects overseas. Not only were manufacturers keen to maintain their home and overseas markets, but governments were keen to maintain viable exports. All these financial considerations provided formidable obstructions to bans on persistent organochlorine insecticides, and they were reinforced by personal and organisational considerations. Personal motivations, involving (as they did) both reputations and livelihoods, were extremely strong – perhaps even stronger than the economic and political motivations. Many research workers and advisers had made their reputations by inventing or promoting the use of persistent organochlorine insecticides. Many took great satisfaction from their success in saving lives and increasing crop yields. Naturally they did not like it when we provided evidence that the persistent organochlorine insecticides were not as totally beneficial as they had believed them to be. However much we acknowledged the benefits of using these chemicals and we were conscientious in doing this – we appeared to oppose progress and to value Peregrine Falcons more than people. This gibe was actually made on more than one occasion!

We were also up against organisational territorial behaviour. I have always been fascinated by the territorial behaviour of both animals and men. The relationship between individual animals with particular places always seems to me to be one of the fundamental facts of biology, and, until recently, it has not received the attention it deserves. My own scientific work on the subject has been on the territory of dragonflies. Territorial behaviour occurs in numerous animals – in fiddler crabs, butterflies, dragonflies, fish, reptiles, birds and mammals. It seems to have several functions and these differ from group to group. The sociobiological approach of recent years, which relates the behavioural aspects of zoology with genetic and evolutionary ones, has given us much insight into animal territory. Its fundamental nature is now better appreciated and understood, and though it is dangerous to extrapolate from one group of animals to another, few would deny that the study of territorial behaviour in animals is relevant to the study of man.

Introspection tells us how deeply we, as individuals, are attached to the places in which we live or over which we feel we have claims. Our emotional reaction to trespass seems to exceed any rational appraisal of likely damage to ourselves or property by the trespasser. Good fences do indeed make good neighbours simply by preventing the encounters which lead to territorial behaviour. We extend this feeling of proprietary relationships to favourite seats in pub or church and, more importantly, to abstractions. In particular, individual people are intensely territorial about their own roles in society and they transfer this territorial behaviour to the organisations to which they belong. Anyone who has worked in an organisation which is subdivided into departments will be aware of the strength of the territorial behaviour of individuals on behalf of their departments. Territorial disputes can be about disputed areas, for example over the allocation of rooms in a building, but also about the roles of departments. Similarly the organisations themselves generate territorial behaviour. All too often charitable organisations – conservation ones not excluded – spend much energy on boundary disputes (so do Trades Unions). All this territorial behaviour is essentially about the allocation of resources. Seeing how important territory is to animals, we should not be surprised that its analogues in human society are also very important. I suspect that we underestimate the extent that territorial behaviour determines the roles and actions of research organisations.

I have digressed on the subject of territorial behaviour because one of the main obstacles which confronted us at Monks Wood was the opposition of other laboratories and organisations. The members of at least one agricultural research laboratory felt that we were trespassing on their territory. We were studying the effects of agricultural technology on agricultural land. They, not we, should have been doing the work. In fact, they did do some, but not enough. Of course, some members of this and other agricultural laboratories always took a more constructive view, and today far better relations exist between the laboratories concerned. But during those crucial years in the 1960s we had to deal not only with opposing vested interests but with the territorial behaviour of laboratories who felt threatened by our work. We had a tough row to hoe.

Whatever their motivation, those who wished to oppose our viewpoint had some powerful instruments to hand and, naturally, they used them. The first was scientific methodology. As mentioned earlier, the species which had suffered most from pesticides were so rare that it was quite impossible on conservation grounds to collect any specimens for chemical analysis; we had to make do with corpses found. We had to admit that these might give biased information; secondly, it was impossible to carry out field experiments on predatory birds since the resources were not available. Therefore, if we were to obtain action before it was too late, we always had to base our case on information which, by its nature, could not be conclusive.

In fact, the agricultural authorities rarely demanded of themselves the level of proof which they demanded of us. Agricultural procedures, including the extensive use of new pesticides, were often adopted without conclusive proof that they were effective.

Nevertheless the methodological shortcomings of our work were real, and were as apparent to us as to any of our critics. There was and is a fundamental division between those who say 'without proper experimentation you can draw no conclusions' and those, like ourselves, who say 'you must draw the best conclusions you can on the evidence available'. Those with the first approach have a good case in academic studies but to us they seemed to be irresponsible in the real world, where the pesticides had already been introduced without fundamental research and were patently doing damage. We felt strongly that if you could not obtain an answer from experiments you had to use circumstantial evidence. This approach is recognised in law and has provided many of the most significant advances in biology – not least Charles Darwin's Theory of Evolution, based on natural selection. In my view, circumstantial evidence is a very powerful scientific tool if used with extreme rigour. We had to use it whenever we could not perform conclusive toxicological or field experiments.

No one could dispute the widespread occurrence of pesticide residues which we found, nor their presence in specimens of species which had undoubtedly declined. Careful use of all the information which 'stood round' these facts demonstrated that the details of the declines fitted exactly with the use of the chemicals and not with other factors which theoretically could be advanced as possible causes of the declines. Detailed knowledge of the 'minute particulars' was essential if circumstantial evidence was to be used. For example, a generalist opposing the thesis that pesticides had caused the decline of the Peregrine might advance the argument that bird populations frequently fluctuate, and the decline of the Peregrine was simply a natural one which coincided with the use of persistent organochlorine insecticides. The generalisation was true: many bird species do fluctuate but it was not applicable in this particular case because there is much information to show that the Peregrine population had been remarkably stable before the introduction of the pesticides. Therefore its decline was likely to be due to an entirely new factor. Studies on all the possible causes showed that only the use of the persistent organochlorine insecticides fitted the decline exactly in space and time.

When some of our opponents were finally convinced by the logic of our arguments that these chemicals were at least very damaging to some birds of prey they often shifted their ground and said 'Yes, it's sad that these birds are going but does it really matter; surely their survival is outweighed by the huge benefits brought by the pesticides?' This could degenerate into the accusation that we cared more for Peregrines than people. If our only concern had been for birds of prey we could have been rightly accused of devoting too much time and resources on them, but we felt that they were merely the tip of the iceberg. If the use of persistent organochlorine insecticides continued to increase, many more species would be at risk. In other words, the birds of prey were only indicators; they were the canaries in the mines. The main issue was not their survival, but the problem which they illuminated. The significance of the effects on the birds of prey was only properly appreciated when linked with our discovery that these particular insecticides were found everywhere. In 1965 they were even found in penguins in the Antarctic. During the years which preceded the published reviews on the persistent organochlorine insecticides (see pp. 192/3) I found that our work on birds of prey did more than anything else to alert people to the general problem, but eventually it was the demonstration that pesticide residues were universal which did most to provoke action. It cut more ice with sceptics because the evidence for it could not be challenged seriously and could not be considered trivial. However, both types of evidence were related and were crucial and, eventually, the right conclusions were drawn from them.

I must stress the word 'eventually' because it took nearly 20 years to achieve all the restrictions which we felt were necessary. As explained earlier, the first (and, in many ways, the most important) restriction that on the use of aldrin, dieldrin and heptachlor as cereal seed dressing on spring corn had been achieved in 1961 with relatively little difficulty, and with no more research than the recording of incidents in which birds and mammals had died of direct poisoning. During 1963 the Advisory Committee carried out its first Review of the Persistent Organochlorine Pesticides and published the results and recommendations in February 1964. We had made all our published and unpublished data available to the Advisory Committee and had recommended the complete withdrawal of aldrin, dieldrin and heptachlor and the phasing out of DDT. After days of debate the final conclusions on the environmental aspects of the problem were written with great caution in the published report. The relevant paragraphs stated:

131. On hazards to wild life, we are satisfied that the restrictions placed on the use of aldrin, dieldrin and heptachlor in cereal seed dressings in 1961 are serving their purpose, and have very greatly reduced the number of deaths of seed-eating birds through these chemicals.

132. We accept that some bird deaths, probably due to persistent organochlorine pesticides, may still be occurring which cannot be attributed to seed dressings. Although little is yet known about the toxicological significance of the residue levels found in birds, we agree that there is circumstantial evidence for the view that the decline in populations of certain predatory birds is related to the residues found in such species arising from the use of aldrin, dieldrin and heptachlor and, to some extent, DDT. We have received no evidence that the populations of other species have been affected by pesticides.

133. Residues of persistent organochlorine pesticides found in birds' eggs are in most cases very small. Eggs containing these residues have, however, been found in widely separated parts of the country, and in a few cases the residues were substantial in amount. They may have an adverse effect on egg hatchability, but there is little experimental evidence on which to assess the significance of given residues.

134. No decline in the populations of garden birds came to our notice. Residues of organochlorine pesticides have been found in dead birds (old and very young) and eggs taken from gardens, but we have found no definite evidence to show that these residues were the result of the garden use of these pesticides. Nevertheless, although the garden use of these pesticides is relatively very small, discontinuance of the use of certain persistent organochlorine pesticides in gardens would be in accord with any general concept of reducing the total environmental contamination whenever satisfactory alternatives are available.

135. Although a degree of persistence is desirable for most pesticidal preparations, we take the view that the pesticides which are used should be no more persistent than is necessary for effective control; should be of the lowest possible toxicity to other species; and should not be used more widely than is necessary to achieve their purpose. We are firmly of the opinion that the present accumulative contamination of the environment by the more persistent organochlorine pesticides should be curtailed. Having reached this conclusion, we have sought an order of possible priorities by which their total usage in agriculture, horticulture and food storage practice could be reduced without serious set-back to pest control.

136. Aldrinated fertilizers are often applied annually as a form of insurance – this being an economical way of combating wireworm in potatoes at the same time as the farmer applies fertilizer. Such annual dosage is quite unnecessary for adequate pest control, and the availability of aldrin in fertilizer encourages the farmer to put aldrin on his land more often than is needed. For these reasons we are firmly of the opinion that the use of aldrin in fertilizer mixtures should be discontinued.

141. (part) We hope that efforts will be made to find equally effective, but less persistent, pesticides to replace the more persistent DDT.

And, finally:

143. The accumulative contamination of an environment by persistent pesticides from all sources is a factor which should be given greater weight by all concerned in proposals for the safe use of such chemicals.

The government accepted the recommendations and, as a result, many uses of aldrin, dieldrin and heptachlor were phased out and the use of DDT was to be reviewed at the end of 3 years. Perhaps even more important, the dangers of persistence had been recognised. We had not got all we asked for, but we had got a great deal. The restrictions on aldrinated fertilisers and on the use of dieldrin as a sheep dip were particularly valuable. The latter being the cause of breeding failure among Golden Eagles and of serious contamination of fresh water and estuaries.

By the time the next report – the Further Review of Certain Persistent Organochlorine Pesticides Used in Great Britain – was published in 1969, we had accumulated far more evidence about the contamination of the environment by these compounds. There was much more information about their toxicological effects and there was evidence that the restrictions on the pesticides which had already been made were enabling the affected species to recover. We felt that our proposals to ban aldrin and dieldrin completely and to phase out DDT had even firmer support and we were sadly disillusioned with the weakness of the 1969 report and its recommendations when it finally appeared. True it put restrictions on all remaining uses of aldrin, dieldrin, heptachlor, endrin, DDT and TDE (except for listed exceptions), but the concessions seemed to have been made grudgingly. At Monks Wood we felt that the document was a poor reward for so much hard work. However, from my contacts with Ministry officials and with the chemical industry, I knew that the real situation was better than it appeared.

The 1969 report had elements of a rearguard action and was somewhat of a face-saver for those who had opposed us. We had been much more successful in getting a change of heart than had appeared, and public opinion both in Britain and abroad would never allow a reversal of policy. The persistent organochlorine insecticides were on the way out – the only question was how long it would take to get rid of the remaining uses. Of these, that of the autumn use of aldrin and dieldrin as seed dressings was the most important. We got increasing evidence that delays in sowing autumn corn dressed with these insecticides caused it to be sown early in the new year rather than the autumn and hence at a time when wildlife was much more vulnerable. This caused quite numerous casualties in some years. However, it was not until 1973 that all supplies of aldrin and dieldrin for cereal seed dressing had to cease. Dieldrin was finally banned totally in 1981 and DDT in 1982, except for emergency use against cutworms.

Persistent organochlorine insecticides had had very significant rôles in the agricultural scene for about 35 years. They had played an effective part in raising crop yields, but the evidence on the harm they did to the environment was strong enough to cause their withdrawal before resistance to them became important and before any species had been exterminated by them. While we felt that our research had not been adequately used by the 1969 Review on Persistent Organochlorine Pesticides, that year saw its application in the industrial field to an extent which surprised and heartened us.

17

*

Guillemots and industry

The chemical analysis by gas/liquid chromatography of many of our specimens of wildlife revealed peaks on the chromatogram which could not be attributed to organochlorine insecticides. We did not know what chemical substance they represented. In 1966 we learnt what they were from a paper by Dr S. Jensen of Sweden. He had shown that they indicated the presence of a group of industrial pollutants known as polychlorinated biphenyls (PCBs). These substances were used in paints, plasticisers, waterproof sealers, printing inks, synthetic adhesives, hydraulic fluids, thermostats, cutting oils, grinding fluids and in electrical transformers. They could become widely dispersed in the environment in industrial smoke, from the exhaust of aircraft engines and the hulls of yachts and as factory effluent.

PCBs are chemically related to pesticides like DDT and, like DDT, are soluble in fat. Despite our small resources I felt that it was essential to study these chemicals, not least because effects which we had attributed to organochlorine insecticides might conceivably be due to PCBs. Examination of specimens obtained in our studies of the distribution of organochlorine insecticides showed that PCBs were widely distributed in birds in the terrestrial, fresh water and marine environments of Great Britain. Papers soon appeared showing that PCBs were widely distributed not only in the Swedish and British environments but also in the Netherlands and the USA.

Almost nothing was known about the toxicity of PCBs to birds so Don Jefferies carried out reconnaissance studies on the effects of one form of PCB (Arochlor 1254) on Bengalese Finches. He showed that it only had about 1/13 of the toxicity of DDT. However, other forms could be more toxic to birds than DDT. Clearly they could be a cause of death in the field. Our findings gave considerable support to our warnings about the dangers of fat-soluble persistent compounds. Hitherto, the warnings had been theoretical; now we had evidence that another group of chemicals with these same characteristics was indeed widespread and could be affecting animals over large parts of the globe.

To their considerable credit, these warnings had been heeded by Monsanto, the firm which enjoyed a virtual monopoly in the manufacture of PCBs in North America and much of Europe. Monsanto had seen a paper which Ian Prestt, Don Jefferies and I had published in the first number of the new journal *Environmental Pollution*. This journal had been founded and was edited by Professor Kenneth Mellanby, the Director of Monks Wood Experimental Station. Monsanto quickly got in touch with me and asked if their representatives from USA, Europe and Britain could have a discussion with those of us who were working on PCBs. The meeting was held at Monks Wood on 19 May 1970. We were expecting it to be a difficult occasion. In the event, it was as amicable as it was scientifically interesting. The Monsanto representatives told us about their meetings with others working on PCBs in USA, Canada and Europe. We discussed what was known about the distribution and toxicity of PCBs, the difficulties of chemical analysis and other topics. We told them what we knew. The Monsanto representatives fully understood the close parallel with DDT, and were determined to control emissions of PCBs into the environment themselves, before being forced to do so by legislation. They were frank about the need to be seen to do what was environmentally right. They made it clear that they would seek to restrict the sale of PCBs in the future to those uses which could not cause environmental pollution. We had another meeting in November and there were further discussions with our parent body, the Natural Environment Research Council. Monsanto was as good as its word and the promised restrictions were soon put into practice.

The problem of environmental pollution by PCBs was not, and is not, wholly solved, because other manufacturers (notably those in eastern Europe) still make PCBs available for uses which can cause environmental contamination, and so PCBs remain a worldwide contaminant. The amounts found in seals and sea birds in the Baltic were often considerable, and there was some evidence that they were having significant biological effects in that region. In 1978 I hoped to learn something about the use of PCBs in the Soviet Union as I was invited to attend an international symposium on Global Integrated Monitoring of Environmental Pollution. It was organised by the World Meteorological Organization and the United Nations Environmental Program and was held in Riga in Latvia. The long, slow train journey from Moscow to Riga provided us with a fascinating transect of western Russia; stops in remote village railway stations were long enough to let us stamp our feet in the snow, but we were grateful to get back to the train with its samovars in every coach and Georgian brandy with our meals. As soon as I could, I went a walk along the frozen Baltic shore and wondered about the extent and origin of its pollution by PCBs. We tried to find out at the conference but, if the Russians knew, they were not giving us any information.

The next spring I was lecturing at the University of Uppsala in Sweden, and here I learnt first hand about the studies of Dr Kihlström and his colleagues. His toxicological work on mink showed that the reproduction of these animals could be affected by PCBs. It had also been discovered that Baltic seals with larger than average residues of PCBs had more frequently absorbed their embryos than those with less PCB. This suggested that the decline of seals in the Baltic might be due, at least in part, to contamination by PCBs.

The extent to which PCBs have affected British wildlife has not been evaluated. However, one of the events which may have influenced Monsanto's decision to restrict the uses of PCBs in 1970 was the sea bird disaster in 1969. The great sea bird colonies of the British Isles are internationally important and, because many are easily accessible, they are enjoyed by thousands of people who visit them each year from home and abroad. Therefore there is always much concern if anything does damage to this important part of our national heritage. In the autumn of 1969 it soon became obvious that something was seriously affecting the Guillemot population in the Irish Sea (Fig. 43). Between September and November nearly 17 000 Guillemot corpses had been washed up on the shore. Many more must have sunk to the bottom of the sea before reaching the shore. No one knows how many Guillemots died, but it could have been as much as 5% of the total population. There was very real cause for concern. Chemical analyses of birds picked up dead and of some healthy birds shot for comparison did not show very great differences between them; however, the birds picked up dead did contain more PCBs and DDE in their livers than did the shot birds. The former had lost much of their fat, and we inferred from this that the higher levels of PCBs and DDE in their livers was due to mobilisation of their fat. Eventually it was possible to analyse a larger sample of birds and it was found that the birds picked up dead on average contained about twice as much PCB as the healthy birds.

Fig. 43. Guillemots (*Uria aalge*). Over 17 000 Guillemots died in the Irish Sea disaster in 1969. Their bodies contained more PCB (an industrial pollutant) than usual. Following the disaster, measures to control the amount of PCB entering the environment were introduced by their manufacturer.

We may never know the exact cause of the 1969 sea bird disaster, but it was probably due to a combination of factors. The most likely explanation was this. Dumping or an industrial accident produced unusually high concentrations of PCBs in the Irish Sea, and as a result the residues of PCBs increased in the fat of the Guillemots. Then autumn gales, possibly exacerbated by a shortage of fish in the area, caused starvation and hence the mobilisation of fat reserves, which released PCBs into the blood streams of the Guillemots and killed them.

Many of us had little doubt that the role of PCBs had been significant and that we should publicise the event as yet another example of a persistent fat-soluble substance providing an actual or potential threat to wildlife. The 'official' line was more cautious: it was felt that since we could not prove that PCBs had had a significant effect on the Guillemots it was scaremongering to suggest that PCBs might be involved. There were parallels with pesticides: 'scientific purity' was used as an excuse not to take the action demanded by a commonsense appraisal of the total situation. Fortunately in this case commonsense prevailed among those who could do most about it – the manufacturers of PCBs.

18

Zinc smelting and Brussels

PCBs and the organochlorine insecticides are organic substances based on carbon and chlorine. They are eventually
broken down by organisms and the action of the sun into
harmless constituents. Where heavy metals occur in pesticides
or industrial effluents they cannot be made harmless in the same
way; once in a particular environment they will remain in it. We
have already seen that damage has been done to wildlife by
fungicides containing the heavy metal mercury. Cadmium is not
used as a pesticide but, increasingly, it is being added to the
environment as a byproduct of industry (notably by zinc smelting and by the use of phosphatic fertiliser, which contains it as
an impurity).

Cadmium pollution is of special interest because its most
serious effects are not likely to be manifest for several years, and
because cadmium has received a considerable amount of international attention. The toxic nature of cadmium has been known
for many years, but it was a disaster that brought it to the public
eye. In 1961, 100 people died of Itai-Itai disease in the Jintsu River
Basin in Honshu, Japan. Itai-Itai means Ouch! Ouch!: the disease
causes severe pain in the bone. The nature of the disease was
demonstrated by three Japanese workers, Hagino, Kobayashi
and Yoshiaka. They showed that the sufferers from the disease
had eaten rice which had been irrigated with water contaminated
by cadmium. The cadmium was derived from metal mines and
smelters in the Jintsu Valley. Fortunately this was an isolated
incident, but there were many indications that cadmium pol-

lution was increasing. Accordingly the EEC instructed a group of us to study the problem and make recommendations.

The Treaty of Rome made no provisions for the environment. Therefore environmental problems within the EEC can only be dealt with in the context 'of a harmonious development of economic activities and a continuous and balanced expansion which constitute the paramount purpose of the Community'. However, in 1973 (following the Stockholm Conference), the EEC did set up 'a programme of action on the environment'. In 1978, the Commission (which is the EEC's bureaucracy), decided to appoint a scientific advisory committee to examine the toxicity and ecotoxicity of chemical compounds. Each nation had to send two delegates. I was one of the two representing the United Kingdom. Our job was to advise the Commission about any industrial substance which they chose to put on the agenda and about the methodology of control. Our operations were made possible by the Sixth Amendment of Directive 67/548/EEC which was concerned with 'the approximation of laws, regulations and administrative provisions relating to the classification, packaging and labelling of dangerous substances'. The 'excuse' to study cadmium was provided by the Swedes who had banned the use of cadmium in some of their products. Sweden was not a member of EEC, but its action could affect trade with the EEC', and this enabled the Commission to ask our committee to advise them about the action which should be taken about cadmium.

We held meetings in Brussels and Luxembourg. Like most people whose work has involved them with international bodies, I developed a love–hate relationship with the organis- ation. My colleagues from the other nations were interesting and warm-hearted people. There was general accord and we never seemed to be plagued by those national antagonisms which rise so easily among political delegates. We did not find it difficult to come to decisions which we could all support.

There was, of course, another side to the coin. The EEC bureaucracy is largely staffed by hard-working men and women, but the system under which they act, or fail to act, is so complicated and so hedged about with administrative con- straints that their labours are usually frustrated. The strange ambivalent nature of the Commission, the impotence of the

European Parliament and the power of national vetoes all combine to make concerted action extremely difficult. To the confirmed European it is all very frustrating. At times there was an element of tragi-comedy. We often worked late and, on one occasion, we all felt that we were too tired to do any homework in our hotels that particular night, so we asked if we could leave our papers on the table in the room where we worked. We were assured that this was in order and that the doors would be locked until our return next morning. I remember leaving a pile of much valued reprints of scientific papers on cadmium effects; one of my continental colleagues even left some confidential documents from his government.

I was the first to return in the morning and, to my surprise, the long table at which we worked was entirely free of paper. I had difficulty in finding an official who could explain what had happened. When at last I found one and asked him where he had put our papers, he replied *'Tout est détruit'*. He seemed quite pleased as if he had scored a point. By now, my continental colleague who had left the confidential papers, had appeared and was even more alarmed than I was. We grilled the functionary and he eventually consented to show us where the waste paper of EEC was stored. We descended deep into the bowels of the huge office block, along endless corridors and passages and eventually came upon a subterranean hall stacked high with the previous day's rubbish. The scene was presided over by two rather solemn Turkish ladies; already sacks were on their way to the furnaces. To their slight surprise my friend and I began to tear open the sacks in search of our papers. It cannot have been an edifying sight as the sacks contained so much more than discarded notes and memoranda. The floor became covered with orange peel and paper handkerchiefs and heaven knows what else. The Turkish ladies took it all surprisingly well. When, by a quite extraordinary stroke of good luck, we found the sack containing our documents and retrieved them, crumpled but intact, they very nearly smiled.

At our meetings in Brussels and Luxembourg we all learnt a good deal about cadmium as well as about the nature of the EEC. Cadmium compounds are present in small but variable amounts in rocks throughout the world, the average concentration in the earth's crust has been estimated at 0.15–0.2 parts per million.

The natural input of cadmium through weathering and erosion has been estimated as 40 tons per year. This amount is insignificant when compared to the input from human activities. The world production of cadmium rose from less than 1000 tons in the period 1910–19 to 152 046 tons in the period 1970–78. This cadmium is used in pigments, stabilisers, electroplating, alloys and batteries. Some of it returns to the environment. However, most of the cadmium in the environment is derived from zinc smelting, the consumption of coal and oil, the production of iron and steel and the use of certain phosphate fertilisers. It has been estimated that 6500 tons of cadmium are put into the environment each year from these sources in the EEC alone. Since cadmium is an element and therefore cannot be broken down into harmless constituents, the cadmium distributed into the environment each year accumulates.

Studies of cadmium residues in soils, plants and animals show a very uneven distribution. Lettuce and spinach are able to absorb more cadmium than cereals and the other vegetables tested. Earthworms and predatory invertebrates contain more than other invertebrates, and individuals of vertebrate species, whose diet includes earthworms, sometimes contain levels of cadmium that suggest that they might be at risk. In the marine environment crabs, molluscs and plankton-feeding birds such as Manx Shearwaters, Fulmars and Puffins contained more than average amounts. However, there was no evidence that any species was being seriously affected by cadmium, although there was evidence that local populations of animals close to smelters etc. were at risk.

It appeared that most animals do not live long enough to accumulate enough cadmium to harm them. By contrast, man is a very long-lived species, and there is some evidence that older people from districts which are more polluted by cadmium than most do show symptoms of cadmium poisoning. The first of these to occur is the presence of proteins of low molecular weight in the urine (proteinuria). Much research needs to be done before the hazard to man can be predicted, but there is a very real possibility that as cadmium levels increase in the environment, an increasing number of old people will suffer from dysfunction of the kidneys. In the case of cadmium, unlike the persistent organochlorine insecticides, man and not wildlife is likely to be the sensitive indicator species of a general hazard! In the circumstances it seemed prudent to take measures to reduce the amount of environmental contamination by cadmium, and we recommended that this should be done.

The cadmium case is an interesting one because there is no immediate danger to man or to the environment and yet, if we continue to pollute the world with cadmium at the present or an increased rate, the time will eventually come when the levels will be high enough to have predictably adverse effects. Some inputs of cadmium into the environment can, and should, be easily reduced. Other inputs are more difficult to reduce or the extraction processes involved are very costly. Doubtless, those with vested interests who do not look beyond the short term will favour inaction, but if the long-term interests of the Community are taken into account ways must be found to slow down the build-up of cadmium in the European environment. The Community ought to take the long-term view: the decision which is eventually taken about this matter will tell us a lot about the nature and effectiveness of the EEC.

19

*

Wider implications

It is interesting to speculate on what would have happened had we not been successful in restricting the use of the persistent organochlorine insecticides. Undoubtedly the raptorial birds which feed on other birds – the Peregrine and Sparrow Hawk and Merlin would have become very rare and probably extinct in the British Isles, and so eventually would the Golden Eagle. There would have been huge reductions among seed- and insect-feeding small birds, pigeons and game birds, foxes and badgers in the arable areas of Britain. In the 1960s it was sometimes suggested that species threatened by pesticides would become resistant: however, the chance was always negligible that rare species with a slow reproduction rate would become resistant *before* insect pests, whose populations were numbered in millions and which quickly reproduced themselves.

Increasing amounts of persistent organochlorines in the sediments of rivers and estuaries would have had increasingly severe effects on the more susceptible species (for example, mayflies and stoneflies, crustacea and fish fry). It would not have been long before the damage done to aquatic ecosystems was reflected in reduced stocks of commercially important fish.

Of course, somewhere along the line, the damage would have become so obvious that restrictions would have been introduced. Our work merely hastened their introduction, and so saved some species from extinction and the country a not inconsiderable sum of money. In Britain the restrictions stopped the mass slaughter of birds and mammals due to eating corn dressed with the more toxic persistent organochlorine insecticides; it allowed the recovery of the species whose populations and/or breeding success had been diminished; perhaps more importantly it made it almost certain that no new fat-soluble persistent pesticide would ever be cleared by the Pesticide Safety Precautions Scheme in the future. The spin off on PCBs has been described.

Sometimes scientists and authorities concerned with the control of pesticides in other countries seemed to pay more attention to our work than those in Britain. When I asked a Soviet biologist why aldrin and dieldrin were not used in the USSR she replied 'We have read your publications about the effects of those chemicals'. One by one, the countries of both western and eastern Europe banned or restricted the use of persistent organochlorine insecticides. There was great interest in our work and I was frequently asked to visit other countries to talk about it. At that time I spoke at seminars or international meetings in France, Sweden, Finland, the Netherlands, Belgium, Switzerland, India, Australia and the USA. Wherever I went I met real concern about the problems posed by persistent chemicals. In most cases I felt it was my job to emphasise the potential hazards of these chemicals but, at the Eleventh Technical meeting of the IUCN at Delhi in 1969, I found myself along with John George of the USA and Mr Alfred Dunbavin Butcher of Australia arguing against a recommendation which if acted upon, would have banned the use of DDT throughout the world. Of course, we were in favour of getting rid of DDT eventually, but a sudden ban on DDT before adequate substitutes had been found would have had disastrous effects on anti-malarial and other public health campaigns. Nearly all the substitutes for DDT posed a greater risk to those who applied the chemicals and were much more expensive. A total ban on DDT could have led both to the collapse of vector-borne disease control and increased poisoning of workers. Some of the delegates at the Delhi Conference really did appear to prefer Peregrine Falcons to human beings.

By and large, the world does seem to be doing the right thing about persistent organochlorine insecticides. The nations of the north, who can afford to take a long view because they are richer and have fewer disease-control problems, have phased out (or are phasing out) the use of these chemicals. In the south, essential uses – for example the spraying of the inside of huts with DDT – continue. However, in some areas in the tropics, the use of some organochlorine insecticides has already been reduced because of their side effects. For example, several countries have learnt the hard way that the use of chemicals like dieldrin on rice may control insect pests at the expense of the fish in the paddy fields. This is an important matter where the fish provide one of the main sources of protein. It is encouraging that the vast World Health Organization (WHO) campaign against river blindness (onchocerciasis) in the Sahel zone of West Africa is using Abate, an organophosphorus insecticide of low toxicity and less persistence than DDT to control the Black Flies which transmit this terrible disease, thus helping to conserve fish stocks in the rivers.

In the long run, resistance of pests and the vectors of disease to particular pesticides make the pesticides ineffective and cause them to be dropped in favour of new compounds. Already there are many areas where DDT can no longer control the mosquitoes which spread malaria and yellow fever: in these places organophosphorus pesticides have to be used. In theory, the onset of resistance can be postponed by ringing the changes and applying a sequence of different pesticides; in practice, this is rarely done either in preventive medicine or in agriculture. In so far as agriculture is concerned, this is partly due to farmers not recognising that the development of resistance to pesticides is virtually inevitable, and so they do not see that measures to postpone it are worthwhile. It is also due to the practical difficulties of organising programmes which effectively ring the changes of pesticide use. Each farm would have to work out its spraying programme in consultation with its neighbours', and a degree of central control by the agricultural departments would be inevitable. Many farmers might consider this too high a price to pay for the benefits obtained. They would prefer to rely on the chemical manufacturers producing new pesticides. However, this is becoming an increasingly difficult task and, eventually, farmers may be forced to take measures against the development of resistance.

The research on persistent organochlorine insecticides had implications which went much further than the control of certain chemicals. Both the research and the attempts to apply it were part of the growing environmental movement and affected its direction. The relationship between the pesticide element and the movement as a whole in the 1960s is worth considering because it underlies the conservation ethos of today. It is also interesting because it demonstrates the power of simple concepts irrespective of their accuracy.

Man-made pollutants were known to be widespread a long time before DDT was used. For years countless holiday-makers had laboriously cleaned off oil from their feet and their beach clothes. People expected the sheep in the Southern Pennines to be a dirty grey colour and smog to envelop our larger towns. Yet few people talked about environmental contamination, or had a concept of worldwide pollution until it was shown that DDT and other persistent organochlorine insecticides could be detected in all parts of the world. Why was this? Partly because the insecticides were much more widely dispersed than other pollutants: there were still oil-free beaches and there were still unpolluted woods where the trees were covered with lichens. Also, it was obvious that insecticides were poisonous or they would not kill insects. Fears on this count were far greater than they need have been, because most people do not understand the basic tenet of toxicology that it is the dose that matters. The layman, despite all evidence to the contrary, divides substances into 'safe' ones and 'poisonous' ones. The toxicologist, on the other hand, works on the assumptions that all substances can be poisonous if taken in large enough doses and that animal and human bodies are adapted to deal with intake of poisons in small amounts. It is the job of the toxicologist to work out what doses of the substance under review have no effect and what doses are harmful. Because people knew that DDT was harmful to insects and could be to man, they thought that all residues of DDT, however small, were dangerous. When this concept was linked with that of pesticides concentrating in food chains the situation did indeed look very alarming.

As mentioned above (p. 181) the simple model of automatic increase of persistent organochlorine insecticides at each level in the food chain was soon shown not to be universally applicable. In other words, predators sometimes contained less pesticide in their bodies than the herbivores on which they fed and, in aquatic animals, uptake through the gills was often – probably usually – more important than uptake from food. Those of us who had to give talks and lectures on the subject frequently emphasised these points yet the simple model of automatic concentration up a food chain was accepted by most people. There was enough truth in it for people not to worry about refinements. It was a powerful, if inaccurate idea and, in so far as persistent pesticides went, it was not too misleading. However, in the minds of many, it became transferred to all pesticides. Even today I frequently meet people who believe that all pesticides become concentrated in food chains. Fortunately the persistent organochlorine insecticides are exceptional (indeed, highly unusual) in this respect.

Despite the inaccuracy of the commonly held view about the concentration of DDT and the other persistent organochlorine insecticides in food chains, it did much to warn the public of the reality and potential dangers of global environmental pollution. DDT became a symbol of pollution. It was obvious that pesticides not only crossed international boundaries in the course of trade but also in ocean currents and in the bodies of fish and birds: pesticides know no frontiers. I used this as a title for a paper I wrote in the *New Scientist* in 1970. The phrase emphasised the international nature of the problem and the need for international action and it tied in closely with the international aspirations of the time.

The value of the DDT symbol in promoting concern about global contamination was to some extent offset by the misconceptions outlined above. It was easy for scientists and regulatory authorities to point out the detailed errors concerning dose rates and concentration factors. These made up the dirty bathwater which accompanied the baby of thoughtful concern about global pollution. All too often both were thrown out. Further, the dangers of DDT were seized upon by those people who wanted to believe that all chemicals were unnatural and bad. By blurring the distinctions between the persistent organochlorine insecticides and the rest, pesticides as a whole could be made to appear suspect. Those of us who were trying to get restrictions on particular chemicals were grateful for general support, but were embarrassed by the excesses of the more extreme environmentalists. It became all too easy for our opponents to lump us all together as the lunatic fringe. I do not think this could have been avoided, but we are paying very heavily for this association. Anyone reading newspapers today or discussing environmental matters with politicians and economists is quickly made aware that the Establishment view belittles environmental problems as such: the only concern is for the effects that environmental issues have on particular business or voting patterns. The environment is still considered subsidiary or trivial and therefore only aesthetes and impractical idealists are concerned about it. Its fundamental relationship with both economics and politics is still not perceived.

The pesticide story had one particular consequence which I believe will be of lasting worth. It demonstrated the value of wildlife as indicators of serious problems. For many years biologists had studied particular organisms as indicator species of general conditions. For example, the distribution of the arrow-worm *Sagitta elegans* (Fig. 44) has been used to determine the extent to which Atlantic water moves up the western approaches of the English Channel. Inshore fishing prospects partly depend upon this influx of enriched water. Earlier we noted the value of the little Bulin snail (*Ena montana*) as an indicator of ancient woodland. Nevertheless the concept of indicator species was one for specialists only. The dramatic decline of the Peregrine in Europe and North America changed all that. People immediately saw that what happened to the Peregrine might happen to animals of economic importance and even to themselves. Thousands of birdwatchers and ornithologists minded very much what happened to the Peregrine, but most people would not have worried unduly if it had become extinct; on the other hand, most people were very much concerned about the possibility of direct or indirect harm to themselves. The Peregrine, in

Fig. 44. Arrow-worm (*Sagitta elegans*). A planktonic animal used as an indicator of the extent of Atlantic water in the approaches of the English Channel (enlarged).

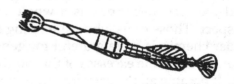

historical times at least, has always had a limited distribution in Britain so that relatively few people saw them, but when common birds like the Sparrow Hawk and Kestrel disappeared over hundreds of square miles many more people were made aware of the pesticide problem through these indicator species.

I think it is true to say that, in future, naturalists and biologists will be more on the alert and more concerned to discover the reasons for the declines of the species which they observe. Experience with the birds of prey showed that conspicuous species can decline catastrophically and yet no one be aware of the fact until thorough surveys have been made. Indicator species have no value unless workers in the field are on the alert and looking for them. This suggests that we should monitor changes in the distribution and status of as many species as possible. Unfortunately monitoring is expensive in time and money and so we have to try and select likely indicator species. Since the chance of selecting the right ones is rather small we shall always have to rely on sensitive serendipity to a large extent. The whole naturalist movement can provide a national service of great value by keeping a vigilant eye on the whole countryside. To conclude, the pesticide saga provided another reason for conserving wildlife – it could warn man of unforeseen danger. The analogy of the miner's canary was a good one.

The pesticide story underlines the importance of time in conservation. It showed how long the environment had to suffer before serious research was done, and how long it took the conclusions of the research to be applied. It also showed how long it took new ideas about a new subject to become common property. The pesticide story also demonstrates that new approaches, new regulations, new actions, in other words, reforms are generated by crisis and disaster as much as by logical thought.

I have already noted that almost as soon as persistent organochlorine insecticides were applied, V. B. Wigglesworth pointed out some of the possible snags in their use. Yet today, 40 years later, farming practice virtually ignores the problem to which Wigglesworth drew attention – the damage to beneficial insects. This is caused by most insecticides and many fungicides and not just by the persistent organochlorine insecticides. Generalisations, however true, do not produce action.

On the other hand, obvious disasters do. The extensive kills of birds and mammals due to eating corn dressed with aldrin, dieldrin and heptachlor in the period 1958–60 resulted in decisive action in 1961; the use of these chemicals on spring sown corn was banned that year. There was no real doubt that the chemicals were killing numerous individuals, including species which were valued as game birds, and that was enough to achieve action. Similarly the extensive restrictions introduced by Monsanto on PCBs in 1970 came within a year of the Irish Sea sea bird disaster, even though there was doubt about the extent to which the disaster had been caused by PCBs.

Nevertheless, the policy of phasing out all the persistent organochlorine insecticides was explicitly based on logical inference – on the worldwide occurrence of the chemicals, on their known toxicity and on their potential threat to the environment. It was initiated by the Advisory Committee on Poisonous Substances Used in Agriculture and Food Storage (later the Advisory Committee on Pesticides) in 1964, that is about 20 years after persistent organochlorine insecticides became available for use in agriculture, and about 3 years after concern about environmental contamination by these chemicals had been expressed. However, I think it is fair to ask whether logic would have been enough on its own. I suspect that the official mind was concentrated by the catastrophic declines of the birds of prey and the public outcry which they produced.

The follow up of the 1964 decision was painfully slow. This was due to interacting factors. First, the law of diminishing returns: the bans achieved from 1961 to 1965, particularly those on cereal seed dressings and sheep dip, had greatly reduced hazards to wildlife, and our monitoring programmes showed a general decline of dieldrin in the environment. This reduced concern and hence pressure on the government and its official committees. In turn this reduced pressure on the chemical industry to find alternatives to persistent organochlorine insecticides and, so long as there were no alternatives, the Advisory Committee was unwilling to restrict uses on particular crops. For export reasons the chemical firms did not want total bans on their products in the United Kingdom. There were even demands for the lifting of some of the restrictions, notably on the use of dieldrin sheep dip. Despite these pressures the line was held by the Further Review of Certain Persistent Organochlorine Pesticides used in Great Britain in 1969. At the time it seemed a great disappointment to those of us who had provided so much more information about the harm these chemicals had on wildlife, but it confirmed that there was to be no going back, and by 1983 all regular uses of the persistent organochlorine insecticides ceased. Thus it had taken 20 years to complete the reforms for which we had pressed in the early 1960s. Logic had won in the end, step by step, but the most important steps were triggered off by disasters.

The crucial problem posed by time was this. If we had waited until we had absolute proof that the persistent organochlorine insecticides were damaging populations, the delay would almost certainly have been enough to cause the extinction of the species most affected. In other words, political action had to be taken on inadequate scientific evidence. Those of us who advocated restrictions on the chemicals were therefore vulnerable to accusations that we were being unscientific. Scientists are indeed divided into two groups – those who will not advise action until all possible evidence is obtained, and those who are prepared to advise on the strength of the available evidence. Whether one belongs to one group or another depends on the value one puts on the action which is to follow the advice. Scientists of the Nature Conservancy are automatically put into the second group by their statutory duty to conserve. We never denied the deficiencies in our knowledge, but we felt that all things considered, there was enough evidence to restrict the use of the persistent organochlorine insecticides.

Scientists of the first group believe that conclusions can only be based on well-designed experiments. Scientists of the second group, while agreeing that experimentation should always be done where it is possible, believe that it is essential to take circumstantial evidence into account. We believe that circumstantial evidence is less valuable than evidence based on experiment but that, if it is used rigorously, it can provide the basis for decision making when the action taken cannot be deferred. Of course, all scientists use circumstantial evidence in practice, but we live at a time when it is unfashionable to admit it. The situation which we faced with the persistent organochlorine insecticides will recur many times in the future. It is to be hoped that those who have to make decisions will take a more rigorous view of the nature of evidence than was taken by some of those who opposed the restrictions on organochlorine insecticides.

Effectiveness of reform depends upon acting in time. The restrictions on the persistent organochlorine insecticides did come in time to save the indicators. The Peregrine has returned to nearly all of its old haunts. Sparrow Hawks have returned to most of theirs, although they are still rare in many parts of East Anglia. The Kestrel is again a common bird in that part of England. These cases should remind us that the margin of error is small. It is better for governments to make occasional mistakes and err on the side of safety than to risk disaster by demanding absolute proof before action is taken. Fortunately the restrictions on the organochlorine insecticides were made before major (and possibly irreparable) damage was done to the fauna of Britain and its surrounding seas.

The PCB story shows that the lesson of the persistent organochlorine insecticides had been learnt as regards one other persistent contaminant. The reaction of the EEC to the advice given it about cadmium will show whether nations can work together to forestall pollution dangers which will afflict other generations rather than our own.

Our research on the side effects of pesticides involved chemistry, toxicology, entomology, ornithology, botany and ecology: we had to operate as a multidisciplinary team.We were one of the first, and they are still so unusual that it is worth describing what we learnt from success and failure.

We found that close working links between the toxicologists and biologists with the chemists were particularly valuable. For example, both sides came to understand the constraints of sample size (both chemical and biological) and hence what was feasible. Similarly the links between toxicologists and field biologists ensured relevance and realism in what each group was trying to do. Many effective connections were made by all of us simply because our laboratories were very close to each other and because we formed a single administrative unit. However, we never achieved one research programme that involved all the members of the team. As mentioned in Chapter 14, I had had hopes of studying the total effects of pesticides on the ecosystem of a whole farm but, for political reasons, the Nature Conservancy was unable to acquire a farm for this important work, and so the project had to be dropped since to base so large an enterprise on land without security of tenure would have been irresponsible. Nevertheless, our team work produced results which could not have been achieved in any other way, and it showed great potential. In 1973 the research branch of the Nature Conservancy, of which Monks Wood was a part, was separated from the conservation branch. It remained in the Natural Environment Research Council but was renamed the Institute of Terrestrial Ecology (ITE). One of the first actions of ITE was to disband the multidisciplinary teams and revert to an organisation based on zoology, botany and chemistry departments. I believe that this was a mistaken policy and that ITE lost a great opportunity in not building on the multidisciplinary foundations which we had laid.

At the time of writing, environmental research is being subjected to savage governmental cuts. They are being implemented piecemeal and this obscures their overall significance, which is no less than a major reduction in government support to environmental science. If ITE had consisted of a few viable multidisciplinary teams doing obviously valuable work on major topics, it would have been much easier to defend today. So ITE in particular, and science in general, is suffering from the failure to perceive the need for multidisciplinary organisation.

Multidisciplinary work depends on effective contact between research workers. Therefore the size of the multidisciplinary research section or unit is limiting. At its largest the Toxic Chemicals and Wildlife Section never consisted of more than nineteen scientists and eleven assistants. I suspect that if our section had been much larger we could not have functioned as an operational unit. We were most fortunate in being the right-sized unit in the right-sized research station, since Monks Wood's staff never exceeded 110 people. Thanks to Professor Kenneth Mellanby's imaginative direction of Monks Wood, we were given admirably free conditions under which to work. If the station had been much larger more bureaucracy would have been inevitable and enthusiasm (and hence efficiency) would have been reduced. Large size is bound to provide problems; on the other hand units can be too small to be viable. Hence, medium-sized research units and stations appear to be the best for environmental research.

Our work at Monks Wood was concerned with both crops and wildlife – agriculture and conservation: as so often happens in science, particularly interesting things were happening on the boundaries between subjects and between disciplines. Unfortunately, the way that science is organised hinders rather than helps research in the areas where there is an overlap. The Agriculture and Food Research Council (AFRC, formerly ARC) looks after agricultural research, the Natural Environment Research Council (NERC) environmental research. When I served on the committee which advises NERC on the allocation of research grants I soon learnt that proposals which had agricultural elements would be disposed of by sending them to AFRC and, likewise, AFRC would pass agricultural projects with environmental components to NERC. As a result, hybrid projects mostly fell between two stools and were not supported by either research council as they were partly outside their respective remits. Some way must be found to get round this purely organisational problem in the future.

Research on the environment is bound to involve different disciplines and hence people with different training and basic assumptions, therefore communication is crucial. At Monks Wood we showed that effective cooperation was possible not only between scientists of different disciplines but between research scientists and the regional staff of the Nature Conservancy, who had to manage the nature reserves and select the SSSI. We had a long way still to go but I believe we were on the right lines. The separation of the research functions from the conservation functions of the Nature Conservancy in 1973 was done for political reasons and in accord with the customer/contractor principle of the Rothschild Report. In my view, this was a retrograde step which wasted public money and expertise and was bad both for conservation and research, as was the break up of the multidisciplinary teams. Much of this book is about the need to involve the general public in conservation, but that does not mean that the links between conservation and research should be weakened. Quite the contrary, the decline and increasing isolation of habitats will make it even more necessary to base reserve management on good research, and many new problems of biological and chemical pollution will arise and need to be solved. We should strengthen those links between conservation and ecological research, which were such an enlightened feature of the Nature Conservancy in the 1950s and 1960s.

PART IV

*

Towards the future

20

*

Failure to communicate

When the Nature Conservancy was set up in 1949 it was fully recognised that science was necessary for conservation. Experience confirmed this. We could not have acquired all the nature reserves which we have today if we had not been able to demonstrate that habitats were truly under threat. FWAG (see Chapter 8) would not have come into being if we had not shown that hedges and other wildlife habitats on the ordinary farm were being lost at an unprecedented rate. The persistent organochlorine insecticides would have stayed in use much longer and would have done much more damage if we had not demonstrated their widespread occurrence and the harmful effects they had on particular species.

Much of this research was laborious and, in purely scientific terms, unexciting; for these reasons it was difficult to get the necessary resources (both manpower and land were scarce). Shortages caused delay and this meant that action was either in danger of coming too late or had to be based on incomplete evidence. As a result, an immense amount of effort had to be spent in persuading authorities to act on incomplete evidence. In fact, the principal lesson we learnt from experience was that communication with the authorities and the general public was crucial and was much more difficult to achieve than doing the research.

What must we communicate now? We are on the edge of irrevocable change. The changes which will occur if the conservation movement fails will be almost entirely for the worse. Our existence will be immeasurably impoverished. We shall have squandered our inheritance and, as a result, provided immense and totally unnecessary economic burdens on future generations. If conservation is not integrated with development in the ways advocated in the World Conservation Strategy (see p. 249) we shall not only lose the vast resources of the seas and forests but thousands of individual opportunities provided by individual species dependent on the seas and forests. The Great Apes provide a particularly poignant example. By marvellous good fortune two forms of the Gorilla, two species of Chimpanzee, the Orang Utan and seven species of Gibbon have survived into our era. They are much our closest relatives and so we can learn more about the origins of our own nature from these creatures than from any others. Their actual and potential value is immense, to my mind, far greater than any human artefact. Yet all are under threat and the sums of money devoted to their conservation are derisory. The rarest form, the Mountain Gorilla, has a world population of less than 400. In Rwanda it is just holding its own thanks largely to the initiatives of the Fauna and Flora Preservation Society. However, even there, the species is in a precarious position for want of cash. It seems almost incredible that the relatively small sums required to conserve this species and its habitat cannot be found by the international community, but such is the case. It is not an isolated one.

Nature provides marvellous material for television programmes. From early days, conservationists like Sir Peter Scott have used them skilfully to put over the conservation message. The superb series presented by Sir David Attenborough have been seen by millions of people. Clashes between environmentalists and developers also provide good copy for newspapers and periodicals as well as television. As a result, the reading public in countries like the United Kingdom have been exposed for some years to environmental problems. The message has been received and yet the action which should follow its acceptance has been disappointingly meagre. The failure of society to act shows that conservationists are failing to communicate effectively. We must ask why.

Like many applied scientists I started my career believing that the principal block to effective action was lack of scientific knowledge, and that once the true facts were known the appropriate action would follow almost automatically. Experience taught me that in most cases enough was already known to solve the problem. The difficult part was to explain the necessity for action to people with different points of reference and habits of thought, particularly when immediate self interest made them unwilling to try to understand another point of view. By using the same words we kid ourselves into believing that we mean the same thing. We do not! They stand for different things for different people. To the conservationist 'conservation' means looking after resources with overtones of caring and public spiritedness. To many farmers – if it does not mean what they do to grass – the word has overtones of interference and whimsy. To some a 'radical solution' of a problem must almost certainly be the right solution, to others it must almost certainly be the wrong one. Indeed every man or woman is an island with his or her individual symbols. No wonder communication is difficult, when the words we use do not provide a common currency.

Language is not the only block to understanding. The whole framework of ideas on which conservation is based is unfamiliar. The challenge facing mankind is new, therefore it has no cultural roots; it is not enshrined in literature, ancient law or custom. We cannot base action on past experience. To make things worse, it has to be faced when an increasing proportion of the world's population is becoming urban. A city dweller can survive quite happily without pausing to think where the bread and meat come from and what they depend on. The availability of food and drink seems to depend on opening hours and bus services rather than soil, water, plants and animals. Farmers may not use words like habitat and environment but they do know what is needed to produce food. Urbanisation makes it very difficult for modern man to feel his dependence upon nature, even when he accepts the fact intellectually. It cannot be a coincidence that it is precisely those people who have experienced with delight the exuberance and beauty of plants and animals, who have come to understand the fundamental importance of conservation. For many of us, what started as a hobby has evolved into a campaign to persuade our fellow citizens to take conservation seriously.

The environmental problem is a technical, biological one. This adds further to the difficulties of understanding it, because most people have little or no training in biology and many biological concepts are unfamiliar. Most people underestimate the degree to which wild plants and animals depend on particular habitats and they do not think of them in terms of adaptation and evolution. Therefore most do not appreciate the rapidity of technological and social change compared to the slowness of natural selection. Yet the dissonance between the two lies at the root of the environmental problem.

In ordinary life we are mainly concerned with very short time scales. Planning is restricted to today's jobs or, at most, for the term or the year. Just occasionally, as when we choose a career or marry, or take out an insurance policy or draw up a will, we consider long periods within the scale of a human life; apart from making provision for our children and grandchildren we do not raise our sights' further. In settled times, in some parts of the world, landowners have indeed planned for unborn generations, but the rate of change today makes such provision seem almost quixotic. There is very little to make us think about the remote future. Instead we tend to have a static view of things. Trees are treated as if they were rocks or buildings; they are protected by Tree Preservation Orders rather than by replanting. By assuming that the landscape and the habitats of which it is composed are static most people have not been on the look out for the immense environmental changes which have taken place in their lifetimes. Either they have not seen them or have not realised their significance.

Another reason why effective action is not taken is that conservation is considered too trivial to be worth serious support. In the Third World, most people are too preoccupied with surviving in the present to worry about the future. To the unemployed in the relatively prosperous west, conservation (as it is generally conceived today) seems at best irrelevant – at worst a sick joke. If the true significance of conservation is not appreciated it is easy to make it look elitist, yet conservation properly conceived is closely connected both with feeding the world and providing employment.

Some may agree that conservation is important on the world scale, but think that our flora and fauna are so insignificant that it is not worth doing much about them. Indeed why should we take the impoverished habitats in Britain seriously at all? First, they are not totally insignificant: a number of their species have their headquarters in the British Isles – the Grey Seal, the Gannet, the Razorbill of our coasts, the Bluebells in our woods and the Bell Heather in our heaths, to mention just some of the more spectacular. We have an international duty to conserve the habitats of these, not to speak of the estuaries which provide the wintering grounds of numerous wildfowl which breed in other countries. Second, even if Britain represents only a part of the total range of species, it *is* a part, and we should share in collaborative international efforts to maintain the species throughout its range. We never know when our role may not increase in importance. This has already happened in the case of the Golden Eagle and the Peregrine. Fifty years ago, the British populations of these species formed only a small proportion of the total European populations. Since that time both species have declined considerably on the continent, while the British populations are much the same now as they were then. As a result they now form a much larger and very significant proportion of the total European population. This situation may well occur again with other species. Third, and I believe this is far the most important reason, we are citizens of one world and so we should be concerned about world conservation. The best way we can show that we mean business is to do what we can with our own resources on our own doorstep. We cannot expect other nations to heed our advice about conserving the rain forests or halting the desert in the Sahel zone of Africa if we do not practice what we preach at home. And so I believe that the farmer who actively conserves a piece of ancient woodland on his farm or digs a new pond is doing much more than amuse himself or provide a local amenity. There is a link between what he does and major actions designed to save the planet from destruction, because both depend on a belief that conservation matters.

Those who have understood the significance of conservation seek a radical reform. They wish to change the present status quo in which conservation is regarded as peripheral to 'real' life. They want it to become a national objective on a par with the maintenance of peace and the reduction of poverty, not least because they can perceive that maintaining the raw material for future evolution and for future exploitation and enjoyment by mankind has close links with these other great objectives.

However successful conservationists become in communicating the idea that conservation matters, the practice of conservation will always be attended by a very real difficulty which tends to undermine its importance: the total resource is made up of millions of species, but threats to species tend to come one by one. The man in the street is right in thinking that the loss of a single species is often a relatively trivial matter, but the conservationist is right in seeing each species as part of that overall diversity which should be treasured by mankind. The practical consequence of this dilemma is that conservationists frequently find themselves having to fight a series of battles, each of which appears trivial. Therefore it is crucial for conservationists to develop a rigorous sense of proportion. They must get their priorities right and be prepared to admit that some conservation goals do matter more than others. This will often lead them to put more emphasis on the conservation of whole habitats rather than on certain individual species which they may contain.

Action is blocked both by expectations which are too high and by those which are too low. The great advances in understanding the genetic code and the possibilities of genetic engineering encourage people to believe that we are not far from the time when we shall be able to recreate organisms which have become extinct, or at least store the germ cells of those about to become extinct. We can leave it all to science. Of course we cannot. Even if it were possible to recreate the population of a few species in this way, the expense and difficulties in recreating all the thousands of species in a self-perpetuating ecosystem would be prohibitive. In addition I believe that exploration of space and science fiction encourage people to devalue the earth. The earth does seem less valuable if one believes that we could emigrate to another planet and take our plants and animals with us!

I sense that the most serious block to accepting the challenge of a deteriorating environment is the human predicament itself. As the population increases and society becomes more complicated, it is difficult for the individual to believe that he can do anything. This feeling of impotence is made worse by the threat of nuclear war. Indeed nuclear war would undo most if not all previous efforts to conserve; its threat hangs over all of us and, I suspect, has a much more debilitating effect than we like to admit. Many feel helpless or even hopeless. Modern man is deeply perplexed about the nature of existence. Many people find it difficult to base their lives either on belief in a Creator or on unbelief. Doubt about man's fundamental nature affects their view of the future and what they should do about it.

Even belief in a beneficent God can lead to totally opposing conclusions about what to do about the future. Some believers see no need to take action on behalf of the future, because all is in the hands of God and He can be trusted to do what is necessary. Other believers do not take the command of 'take no thought for the morrow' so literally. They see conservation as a form of stewardship to God, which demands action. Similarly, some unbelievers see the world as governed by immense evolving forces and they think that there is nothing that individual men and women can do to oppose the tide of history. Others feel that conservation is a form of stewardship to man. Whatever its philosophical base, fatalism inhibits thought about time and conservation, while stewardship promotes it.

I believe that until people see the significance of conservation by looking at it in terms of time they will not be motivated to demand its practice; until they do, politicians are unlikely to act. Conservation must become political before it can be achieved. Therefore it must be seen to be relevant to people now. Conservation is a new idea, and new ideas always look cranky, simply because they are new and are held by a minority. Conservationists have to be very careful not to appear simple minded and obsessed with their ideas and apparently oblivious of other legitimate aspirations. Conservation is, and should be, promoted as exciting yet also completely sane, down-to-earth and practical. Owing to its newness it is extremely difficult to do this, but it must be attempted because the image of conservation does matter and its promotion cannot be achieved by enthusiasm alone. Advocates of conservation who are accepted as sensible, balanced people have an especially important role to play in promoting conservation. There is no better advocate than a successful farmer who practises the conservation he preaches. Such people can discern the crucial difference between compromise which is merely a cosmetic word to describe defeat by stronger forces, and compromise which achieves effective conservation just because it integrates it with other legitimate interests.

Conservation should be seen to be supported by men and women of good will of all political parties and of none. 'Green' parties may have a role in goading the established political parties to take conservation on board, but it would do conservation much harm if it were thought to be the prerogative of a fringe group. There is room for much political argument about how it should be achieved, but conservation itself should not become identified with one party or caucus. It matters too much.

Conservation must be related much more closely and professionally with the current economic situation and current ways of thinking. Introducing the conservation element into the body politic is a radical matter, and it has to be done at a time when other radical reforms have to be made. In the western world, the old industrial society is being replaced by a new one based on electronics; whatever one's political views there is an obvious need for less rigidity and greater adaptability. The new scene provides great opportunities for conservation, because it could bring employment into areas impoverished by the decline of heavy industry and by their remoteness.

The unfamiliar nature of conservation, its apparent triviality if misunderstood, facile optimism and underlying feelings of helplessness together produce a climate within which it is extremely difficult for conservationists to put their case. Yet, once these blocks to understanding and action have been identified and faced squarely, they can be attacked robustly. This is best done when dealing with particular cases, because whenever there is a conflict of interests the credentials of conservation are bound to be questioned. Over the years conservationists have had experience with numerous particular cases – several have been described in this book. We should have learnt a good deal about tactics.

First, there are conclusions we can draw about the tacticians themselves. Reforms are put into practice by changing attitudes, laws and regulations and by reallocating financial resources. However, they can only be achieved in the first place by individual men and women conducting specific campaigns. Reforms never occur by accident; in every case the inertia of the *status quo* has to be overcome by the positive action of individuals.

Most people are not committed to conservation. If you listen to people talking about conservation you get the general impression that everyone is in favour of it, but when people are faced with the 'minute particulars' – when lifelong habits or the purse are affected – it is given a very low priority. In fact, effective conservation campaigns are achieved by a very small number of committed people. I suspect that all of them identify themselves with the particular campaigns with which they are concerned. As a result they are identified by other people as proponents of the campaigns, and so praise or blame is attached to them according to whether they succeed or fail. They then become yet further motivated and generally more effective. This process of identification can be promoted by a good conservation administrator: he should be able to make a member of his team feel that a certain subject area is his or her territory, and as he or she identifies with it the stronger becomes the commitment. This is the constructive side of human territorial behaviour!

In activities like conservation where the issues are not simple and involve many interests, committees have to be set up to solve problems. Yet anyone who has served on a committee knows that the committee itself never conducts a campaign or introduces a reform. The reform is generated by one or more members of the committee, who then convince their colleagues that the reform is necessary. The committee has great value in enabling responsibility to be shared, but it cannot operate without activists who promote constructive ideas within it.

The success of an activist in a committee depends not only on his or her persuasiveness, but also on the perception of the rest of the committee on how the world outside will react to the reform. To that extent, avowed (though skin deep) approval of society is important. Politicians and all others dependent on votes cannot act without it. The commitment of individual activists is essential, but their success depends largely on getting their timing right, and experience suggests that timing has to be related to the occurrence of disasters.

In previous chapters we have seen how effective disasters were in promoting effective conservation action. For example, the scale and unpleasantness of myxomatosis rapidly produced the administrative machinery to deal with the problem. The scale of bird and mammal casualties due to the spring use of aldrin, dieldrin and heptachlor as cereal seed dressings rapidly produced bans on this use. We also found that the Torrey Canyon incident did more to control major oil spills and to provide remedial treatment of slicks than all the warnings which preceded the disaster. The extinction of a handful of species such as the Large Blue Butterfly has done more to rouse public opinion than the production of all the statistics which demonstrate the massive declines of many surviving species, even though this is a far more important matter. Why are disasters so important a means of communication?

It is the experience of all connected with the media that good news is no news. The conservation bodies have learnt this sad fact again and again. Press notices are frequently issued about new nature reserves or about constructive meetings between conservationists and farmers, but they never make the headlines, whereas hints of conflict or failure are seized upon to make a story. It is almost impossible to get journalists to take up an idea unless there is the peg of a story to hang it on. Therefore if conservationists want to reach a wider public through the media they are forced to hang their ideas on the peg of bad news. A disaster is a potent type of bad news for it engages the attention and leads to questions about its cause and who is to blame. It provides an unusual opportunity for conservationists to put ideas over to the public.

By their nature, disasters occur rapidly and without immediate warning, yet unless conservationists are already prepared they will lose the opportunity to be effective. Most disasters are in fact predictable, the only doubt is about the exact time and place of their occurrence. Therefore conservationists can and should deliberately plan what they want to say and do before a predicted disaster strikes. In this way, disasters can be used constructively to put over ideas and to obtain preventive action. It has to be admitted, though, that the present association in the public mind of disasters and conflict with conservation – the inevitable result of the media concentrating on bad news – has been highly detrimental to the conservation movement. It obscures the fact that most conservation activity is cooperative and positive.

Disasters may occur too soon to be used to carry a message. Public opinion must have advanced enough to receive the new idea. It is easy for a conservationist to be so immersed in his particular cause that he is unaware how unfamiliar it is to other people: it is easy for conservationists to act too far in advance of public opinion. For example, our demonstration of habitat losses in the 1960s was too early to have the immediate effect for which we hoped: in those days most people did not realise that habitat is essential for the survival of species. Human survival is rarely dependent upon the survival of a particular habitat: man can live almost anywhere, and so it was assumed that plants and animals could likewise. However, in recent years, thanks to an increase in the teaching of ecology in schools and universities, and an increase in television programmes about ecology, the concept of habitat and its importance have become familiar; therefore the significance of its loss is now much better understood.

Conservationists face a serious dilemma as regards the timing of their campaigns. If they wait too long before introducing them so much damage will have been done to the resource that their efforts to save it will be too late either for biological reasons or because the action required will be prohibitively expensive to implement. On the other hand, if they act too soon their campaign will be ineffective.

I believe that we have learnt valuable lessons from past experience and that we should be able to apply them to the future, but does our ignorance of the future invalidate this claim?

21

*

Predicting the future

Conservationists cannot predict the future exactly, any more than economists and politicians can. Therefore, is it idle to plan any form of strategy for conservation? I am sure it is not because we can make useful general assumptions about the future even if we cannot predict details. For example, we can be sure that totally novel factors will arise which will have profound effects on conservation, but we cannot predict what they will be. We can be certain that if a particular type of disaster is possible it will almost certainly occur, even if we cannot predict the time and place. We know that numerous species of plants and animals will be of incalculable value to the human race but we cannot predict which those will be. (No one could have predicted that a common species of mould would save millions of lives in this century alone.) These generalisations hold good for the most remote future and we should take them into account in our planning today.

When we consider the immediate future we can make more detailed assumptions, and so can consider the options which appear to exist. This will enable us to keep one move ahead of events. In Chapter 2 we saw that the original flora and fauna of Britain had been profoundly modified by agricultural and wood-land management for hundreds of years. For the foreseeable future, agriculture and forestry will continue to impose the principal constraints, although they will differ from those of the past. Can agriculture and forestry be modified so that wildlife can live in the crops, or is polarisation into land for agriculture and forestry on the one hand, and land for conservation on the other inevitable?

Agricultural and forestry policy will be crucial. Agriculture and forestry are constrained by economic, financial and political factors. The situation which results is so complicated that there is no way of predicting how it will evolve. Quite small shifts in accountancy can have immense effects on the ground. Oil seed rape provides a striking example. In order to reduce imports of vegetable oils the EEC has encouraged the growing of oil seed rape. As a result, much of eastern England is now bright yellow in May. Oil seed rape is attacked by pollen and seed beetles which greatly reduce yields. Accordingly, the chemical most effective in controlling these pests – the organophosphorus insecticide Triazophos – is widely used. If applied according to the instructions on the label it does little harm to bees, but if it is sprayed when some of the crop is still in flower, it can have a devastating effect. In 1981 about 0.6% of the hives in oil seed rape growing areas were affected by this insecticide. The problem for the farmer is real. Small differences in soil and aspect can bring different parts of the crop in a field into flower at different times. Sometimes it is almost impossible to spray without affecting some bees. We can measure the effect on bees, but no measurements have been made of the effects on the numerous wild insects which also feed on oil seed rape. In a landscape in which most hedges (and hence wild flowers) have been removed, oil seed rape is a very significant source of food for bees and other insects. Thus that simple EEC decision in Brussels has transformed the appearance of much of the English countryside. It has had unforeseen effects on the bee-keeping industry and unmeasured, but doubtless considerable effects on numerous species of wild insects.

Apart from atomic warfare, nothing could affect conservation more in the future than agricultural policy. Lack of government support for agriculture in the 1920s and 1930s prolonged the existence of traditional farming into those decades, and so postponed the clash with conservation that was bound to occur once technological information was applied extensively to agriculture. If government withdrew support for agriculture now many of the pressures on marginal land would eventually decline, although the immediate response might be an increase in the destruction of wildlife habitats as farmers tried to meet new costs. It is certain that the countryside could not return to the way it was before the Second World War: ancient woods and hedges which had been lost during and since the war could not be recreated, even if there were the will to do so; poor drainage would increase the numbers of wetlands and derelict grass leys would gradually become floristically richer, but the sources of seed are now so much reduced that it would take many years before pastures had any resemblance to those which existed before the war.

This point is often underestimated by those who claim that new habitats such as motorway verges and disused quarries will go a long way in compensating for loss of habitats on farms. Experience on my own doorstep has shown me how slow plants are to disperse. I am lucky enough to own the field behind my house and garden. Twenty years ago I set aside an acre of this field as a nature reserve. I have excluded grazing animals from the area and planted about 1000 trees belonging to 20 native species. Numerous invertebrates and some bird species have colonised the reserve, but not a single species of plant, even though digging and mowing have produced a wide range of suitable conditions for colonists. My field is 3 km (2 miles) from the nearest wood and about 1.6 km (1 mile) from the nearest old pasture. It is surrounded by gardens and intensively farmed arable land. It seems that this degree of isolation is enough to make colonisation an extremely slow process for plants.

A drastic decline in the government's support of agriculture would result in a drop in land prices which might encourage the development of numerous small farms, either subsistence farms or part-time farms. Contrary to what is often believed, this would not always be advantageous for wildlife. The small farmer simply cannot afford to leave appreciable areas of his land unfarmed: he will try to cultivate every square yard. Gardens provide a close parallel. The studies by Ian Wyllie of the Cambridgeshire parish of Hilton and my own on the new town of Bar Hill in the same county show that a given area made up of many small gardens produces far less wildlife than the same area made up of a few large ones. There is little room for wildlife habitat in small gardens and disturbance in them is bound to be greater whether it is by people or pets.

I think that we can dismiss scenarios in which agricultural practices are determined solely by market forces, because the chance of future British governments withdrawing all support for agriculture is very slight. The need to produce a large proportion of our temperate food stuffs – the things which we can produce – in order to reduce our import bill will remain. Already Britain exports a considerable amount of food and will probably do so increasingly. Indeed, all the advanced nations in the temperate zone may need to help feed the Third World until population control and improvements in tropical peasant agriculture enable nations in the Third World to feed themselves. Doubtless there will be painful adjustments for the British farmer, but we can be fairly certain that all future governments in the coming decade will give substantial support for agriculture. The painful adjustments will encourage the shedding of yet more labour and, hence, the use of yet more technology to keep the farmer in business. Conservationists would be wise to plan their activities on these assumptions.

We must assume that on the better land the croplands will become less and less suitable for most species of wildlife, and that the latter will depend increasingly on areas deliberately set aside for them, and on forest and recreational land which can be managed secondarily for wildlife. Therefore it appears that, for the foreseeable future, wildlife in Britain will depend upon what will really be a network of public and private nature reserves. At one end of the scale there will be those reserves managed by the Nature Conservancy Council and the voluntary conservation bodies and, at the other, the generally much smaller areas of wood, hedge, grass verge and pond managed by the farmer, with increasing support from FWAG and others concerned with conserving wildlife on ordinary farmland.

Such a scenario is unpopular with many. There are still those who believe that by some manipulation of nitrogen application rates, pesticide use, drainage and the like, an acceptable compromise can be reached which will leave fields both productive and full of wildlife. Of course much damage to wildlife can be prevented by applying fertilisers and pesticides wisely and by avoiding excessive drainage, but there is no evidence whatever to suggest that we can return to a situation where acceptable productivity and wildlife conservation can take place on the croplands themselves. This is because we need to produce more per acre than we did and therefore the fields no longer provide suitable habitats for most wildlife. Whether land were nationalised, divided up into relatively few large ownerships or divided into numerous small holdings the broad effects would be the same. For the foreseeable future, most wildlife on lowland farms can only be conserved on the land which is not being cropped, in other words in small woods, hedges, ditches and so on. Conservationists should therefore concentrate on how this can be achieved within the framework of productive agriculture. In particular we must relate the functions of nature reserves, SSSI and unscheduled wildlife habitat on the farm much more positively than has been done hitherto.

There is more scope for combining nature conservation with forestry than with agriculture. If deciduous woods are managed on a coppice and standards rotation, practically all conservation requirements are fulfilled. If deciduous woods are managed as high forest the wood lacks the shrub layer but that can be easily provided for by allowing bushes to grow up on the edge of the wood and the rides. Plantations of exotic conifers provide less scope, but it is usually possible to retain some of the original habitat (whether heath, moor or deciduous wood) on the edges of the plantation or on the edges of rides; this can provide a suitable habitat for many of the species dispossessed by the plantation. Of course, some adaptable common species such as the Chaffinch and Carrion Crow will survive in the plantation itself, and species such as Coal Tit and Goldcrest which favour conifers will actually increase because of them. If felling is done at different times a mosaic of habitats is produced and species which thrive at the planting stage but which disappear later, for example Woodlarks and Nightjars, can be provided with suitable habitat on a shifting basis. Also, small modifications in silvicultural treatments (brashing, spraying etc.) can favour wildlife though generally only the commoner species. In general the growing of trees provides much more scope for conservation than the growing of wheat or Italian Rye-grass.

Today the import bill for timber is similar to that for imported food stuffs, yet it is cheaper for Britain to import timber than to produce it. Nevertheless, forestry is given considerable support by the government in order to maintain a strategic reserve in case of emergency: the Forestry Commission is run as a state service and private owners of woodlands receive considerable amounts of grant aid.

Most of the grazing land in upland Britain once supported forest and could do so again. So long as the huge softwood timber resources of Scandinavia, the Soviet Union and North America are available the advantage in growing more of our own softwoods would remain strategic rather than financial. On the other hand, the world's hardwood resources are becoming severely depleted and circumstances could easily arise in which it would make good sense to grow more of our own hardwood. This might be helped by a decline in agricultural support within the EEC since that would free a substantial amount of land for afforestation with hardwoods. Also, eventually, it may even pay to grow quick-growing trees as a source of energy. Taking all these circumstances into account, it is very difficult to predict whether the area of forest land in Britain will remain roughly the same in the near future, or will increase, or even suffer some decline. Conservationists must keep their options open and be prepared to exploit the opportunities that may be provided.

For many centuries wildlife in Britain has survived in an archipelago of habitat islands. In the past most of the islands were much larger. During the period of enclosures (sixteenth to nineteenth centuries) the total forest cover was reduced but links between woods were formed by the planting of hedges. We must keep both habitats and links today. For some species, for example the Black Hairstreak (see p. 16), the existing pattern of habitat no longer allows this. For other species, their ability to colonise new habitats and to recolonise old ones demonstrates that the existing pattern of habitat is adequate for them. The only way we can conserve species like the Black Hairstreak is to ensure their survival in nature reserves by good management and to introduce them into habitats from which they have disappeared. Thus, the greater the reduction of wildlife habitat the greater is the need for sensitive management of nature reserves and for scientific research on the requirements of the species involved.

Thanks to the good offices of their owners and the help provided by the Wildlife and Countryside Act, SSSI will receive an increasing amount of special conservation management. As a result the difference between National Nature Reserves and SSSI will become less. We have to accept that fairly soon the majority of our rarer species will depend on nature reserves and SSSI for their existence. The old county of Cambridgeshire is a good example of a county where agriculture is highly productive and hence agriculture and conservation are necessarily polarised to a large extent. Already a large proportion of this county's rarer species are confined to nature reserves or SSSI within the county – Black-tailed Godwit, Ruff, Bearded Tit, Cetti's Warbler among the birds, Chalkhill Blue, Brown Argus among the butterflies, Juniper, Pasque Flower, Bloody Cranesbill among the plants. Many other species are far more abundant in reserves and SSSI than outside them, for example the Oxlip, Snipe, Redshank, Shoveler and Gadwall. Many other counties are likely to become more like Cambridgeshire in this respect.

So much for the near future, in which events will prove or disprove these opinions. It is worth considering the remote future also, in case it affects what we should be doing now. Eventually man may be able to synthesise all his food and all the fibres he needs. As a result, all farm and all commercial forest land would be freed for other purposes. The complicated accommodation between farming and conservation, which is so important today, would no longer be necessary, but the basic problem would remain. Nature conservation would still compete with other land users: with housing, industry, recreation and perhaps quite new activities which are totally unforeseen today. As a result, priorities would still have to be determined and sites for conservation selected. Whatever scenario is devised, some land will have to be managed primarily for conservation; on the rest compromises obtained with other land uses. Nature reserves will still be necessary, as will be schemes and organisations to integrate conservation with other uses of land.

All this means that we can build confidently on what already exists. However, any blueprint which we devise should encompass both the immediate future and the more distant. We need both a short-term and a long-term strategy.

22

*

The holding operation required

There is a conservation crisis and, whether we like it or not, today's generations are not only the first to be deeply involved in it, but the only ones which can take effective action. If we fail to do what is required of us, there will be little that future generations can do, because so much of the raw material will have gone for ever. What we ourselves do or do not do about conservation will be crucial. Yet, if we are to be effective we must also be realistic. We have to accept that attitudes are not changed quickly; we cannot rely on a sudden change of heart by the public and hence by politicians. Attitudes are changing. The long-term strategy of education and research is already bearing fruit, but it would be foolhardy to expect dramatic results in the short term. Therefore, running parallel with the long-term strategy we must operate a short-term strategy. Essentially it must be a holding operation. It should cover both the special, protected sites and the wider countryside. We must take the last opportunity that will be given us to establish nature reserves and we must halt the unnecessary destruction of habitats in the wider countryside. We must do these things in our own country and at the same time help other nations to do the same in theirs.

First let us consider the protection of special sites. The objective is to provide bases which support viable populations of as many species as possible. These bases should be numerous enough and close enough together to allow for dispersal between them and to provide sources from which the wider countryside can be colonised when opportunities arise in the future. In deciding how many protected sites there should be, we should take into account the fact that conservation will almost certainly never occur again as the unplanned byproduct of agriculture. The pressures of population and industrialisation are now so great that, unless conscious efforts are made to integrate conservation with other human activities, the decline of habitats will continue almost unabated. In future, conservation will depend largely upon conscious conservation activity. We also have to take into account the fact that we cannot bank on any particular forms of food or fibre production occurring in the future, and hence on any particular degree of compatibility of conservation with agriculture and forestry. We cannot even assume that food and fibre will continue to be produced in fields and forests. We cannot predict climatic change any more than human activities. On the research side we cannot predict how long it will take to obtain the best answers possible to questions about the minimal sizes, numbers and distributional patterns of protected sites. For all these reasons, we must take a very conservative view about sites; we must protect as many as is practical.

In Britain, the existing network of National Nature Reserves, Local Reserves, Forest Reserves, reserves managed by voluntary conservation bodies, and SSSI managed by their owners provide a firm base for the required network (Fig. 45). However, there is an urgent need to implement the programme of reserve acquisition and to notify more SSSI. These requirements are fully supported in the recently published document of the Nature Conservancy Council entitled *Nature Conservation in Great Britain*.

The present selection system of SSSI can be and should be improved, but the main requirement is to speed up the process of notification of these sites. The recent increase in grant-in-aid to the Nature Conservancy Council should go a long way in achieving this. However, it is not enough to cater for the rapid expansion of the programme to establish National Nature Reserves – a thorough programme of site protection is necessary and more money will be needed for that. The sums required are minute when compared to those spent on defence, health, education, and agriculture but would be considerably more than the tiny allowance so far made available for conservation. At present this is equivalent to the cost of a cup of instant coffee per

British citizen per year! If this were raised to the equivalent of the cost of a pint of beer each, Britain would have little difficulty in carrying out an effective programme of site protection.

Conservation is about the present as well as the future: perhaps too little effort has been put into making conservation pay for itself today. Very few organisations charge the public for permission to visit nature reserves. In the case of National Nature Reserves managed by the Nature Conservancy Council this is right because the reserves have been paid for by taxpayers' money, but this objection does not hold with reserves managed by non-governmental societies. Wicken Fen in Cambridgeshire was established as a nature reserve from 1899 onwards and so is one of Britain's oldest nature reserves. Today it is owned by the National Trust. Its very interesting fen flora and fauna depend on mowing and careful maintenance of the dykes. The surrounding land has been well drained and is intensively cultivated and, as a result, the peaty soil has diminished and the level of the land has sunk. Hence Wicken Fen tends to lose water and this encourages the growth of swamp woodland and scrub, which threatened to cover most of the reserve and to exterminate its special fen species. Today fen conditions are being successfully maintained by clearing the scrub and cutting sedge and other types of fen; the old dyke system is being restored and water

Fig. 45. Hartland Moor National Nature Reserve, Dorset. One of the Dorset heathland reserves. But for the efforts of the Nature Conservancy Council, Royal Society for the Protection of Birds and the Dorset Naturalist Trust, this habitat would have virtually disappeared. The most important part of any short-term conservation strategy is to establish nature reserves while it is still possible to do so.

levels manipulated by pumping. All this work costs a considerable amount of money; although the sale of sedge and reed for thatching goes some way towards recouping costs, it is insufficient. Recently the National Trust began to charge those visitors to the Fen who were not members of the National Trust. Few complaints have been received: visitors seem to be happy to contribute to the management of the Fen in this way. Similar arrangements might be made on other reserves which require expensive management. In our commercially orientated society it is no bad thing to remind people that nature reserves can be costly to manage as well as to buy. Effective conservation can rarely be done on the cheap.

There is an additional advantage in drawing attention to reserve management: people can see conservation in action. They can see that although work on reserves can be physically hard it involves many skills and is very interesting. An increasing amount of it is done by volunteers organised by the Nature Conservation Trusts and by the British Trust for Conservation Volunteers, but there is great scope for involving the Manpower Services Commission not only on nature reserves but in managing wildlife habitats in the wider countryside. Pilot schemes have been successful but much more could be done if a greater awareness of the value of habitat management led to a greater support by government for such work. Unemployment would be reduced, conservation and local amenities would benefit directly and the tourist industry indirectly.

We have also to consider the conservation of wildlife habitats in the wider countryside. It will largely depend on reform of EEC's common agricultural policy (CAP), since CAP provides powerful financial inducement to increase farm products at the expense of the environment.

CAP has achieved the aims of European self sufficiency in food and in giving much needed support and security to the farming communities of the member nations. However, these achievements have been obtained at a great cost. The size of the agricultural subsidies have distorted the political and economic framework of EEC. Huge surpluses of food and drink have been produced, which either have to be stored at great expense or sold at uneconomic prices. From the environmental point of view, CAP has been disastrous, since it has encouraged farmers both to reclaim the few remaining habitats on the better land and to reclaim huge areas of marginal land – the latter are just those areas where farming and conservation most require sensible and sensitive integration. The present policy represents a great waste of resources because the potential for high agricultural productivity in the so-called Less Favoured Areas is small, whereas their value for wildlife and landscape is immense. The farming communities in these areas rightly receive financial support for social reasons, but the way this is done – by crude agricultural subsidies – merely helps to increase the surplus of temperate food-stuffs while destroying the capital of landscape and wildlife which provide the safest base for the future prosperity of these areas. By calling them 'less favoured' the question is begged. They are only less favoured from the agricultural point of view. In other ways they are highly favoured, although they require EEC support to use their natural advantages.

The EEC and its constituent governments have taken the easy way out; they have used the simple device of agricultural subsidies to solve complicated social objectives. Of course the problems are real and it is easier to solve them through agricultural support than by integrating conservation, agriculture and forestry. Yet, if integration is not achieved, western Europe will suddenly realise that it has destroyed its natural heritage through ineptitude and laziness, and it is too late or too expensive to repair the damage. Despite many warnings and polemical books such as Marion Shoard's *Theft of the Countryside* and Richard Body's *Agriculture: The Triumph and the Shame*, the public has taken some years to realise that agricultural subsidies when crudely and unimaginatively applied have done more to damage the flora and fauna of Europe than any other single factor.

The remedy is obvious: environmental considerations must be taken into account in the EEC's agricultural and social activities. Either CAP should be reorganised so that conservation is fully integrated with its agricultural and forestry aspects, or it should be replaced by a 'Common Land Resource Policy' which does the same thing under a more appropriate name. In either case, financial support to farmers must include support for conservation as well as for the production of food and fibre.

Although the remedy is obvious, its implementation is difficult. Accounting is difficult because the value of agricultural produce can be determined to a large extent by the market forces acting at any particular time, whereas conservation values cannot be so determined, since they involve the future as well as the present and the future has no purchasing power in the present. This problem is not so very different from that which faces us already when we have to put a financial value on supporting farming communities in the less favoured areas. If governments can achieve financial estimates in support of people, they can also provide them in support of the environments on which people depend.

Another problem which needs much more thought than it has been given so far concerns the time scale of the reform of the CAP. A sudden massive reduction in agricultural support might have the opposite effect to that which some might hope for or expect: those farmers who could do so, might increase their efforts to reclaim marginal land in order to offset the effects of reduced EEC support. The recent introduction of the milk quota may reduce pressure on marginal land, but it will also encourage those farmers who can grow cereals to change from dairy farming to cereal growing – a process which usually leads to the loss of hedges and other wildlife habitats.

The reform of CAP requires close cooperation between agricultural, forestry and conservation interests and organisations and departments. The main obstacle in the path of cooperation, both in CAP and in its constituent countries, is the way we organise ourselves.

In Great Britain our machinery of government is particularly ill-adapted to deal with conservation. Conservation depends on the way land is managed and so it is involved particularly with agriculture and forestry, but also with industry, housing and recreation, and indirectly with planning. At present these activities are the concern of different government departments, which act virtually autonomously. As a result it is extremely difficult to obtain action which depends on the agreement of two or more departments, especially when it requires funding from more than one of them. Conservation suffers particularly in this respect, but it is not unique. The connecting links between education and industry, unemployment and violence, transport and health, to name a few, obviously demand close inter-departmental liaison. They are not getting it. I suspect that many of the difficulties which the United Kingdom experiences today are due to departmentalism and the failure to appreciate connections.

There is no law, regulation or even convention which forces the Ministry of Agriculture and the Forestry Commission on one side and the Nature Conservancy Council on the other to plan or to work together. As a result, the very real difficulties of coordinating conservation, agriculture and forestry on the ground are compounded by the machinery of government which has allowed great gulfs to be fixed between the departments most concerned with land use.

In theory the problem could be solved in either of two ways – better liaison between existing departments or the amalgamation in a new department of all government bodies concerned with land. The operation of the Pesticide Safety Precautions Scheme shows that it is possible to have interdepartmental arrangements which work for the benefit of all the departments involved and hence of the general public. However, the Advisory Committee on Pesticides is only advisory and it does not control finance or carry out actions itself. To be effective, the Agricultural departments and the Department of the Environment would have to have a joint mechanism for spending money on activities which served both agriculture and conservation. So entrenched is our system of accounting departments, and so strong is the territorial behaviour of the existing departments that it seems unlikely that we could achieve such a radical new system of liaison. Therefore the only way by which the machinery of government can be made more effective about conservation would be to abolish the agricultural departments and set up a new 'Department of Land Resources', which covered not only agriculture and forestry but also conservation of natural and archaeological resources and recreation.

I am not recommending that MAFF should take over the Nature Conservancy Council and the Countryside Commission – that arrangement would probably reduce the effectiveness of conservation, since agriculture is much more powerful than conservation. Experience in the USSR, where conservation is dealt with by the agricultural department, confirms this view: whenever there is real conflict between agriculture and conservation, agriculture always seems to come out on top. Food production is, of course, vital and therefore should be the primary land use on most of the land, but that does not mean that whenever there is conflict about a particular place, agriculture should always win. Safeguards to conserve the more important areas for wildlife would have to be built into the system. This would not be difficult since quite refined techniques for assessing both agricultural value and conservation value are already available.

We must conclude that the short-term strategy for conservation within Britain consists of two elements. We must speed up the acquisition of nature reserves and the notification of SSSI and we must use our membership of the EEC to reform the CAP so that it takes account of the environment. All measures would be facilitated by the overhaul of our machinery of government. Other short-term measures are required to promote conservation overseas, and they are equally important.

Of course, reform of the CAP will benefit other members of the EEC as well as ourselves. In Britain we tend to take an insular view of things and forget that the CAP is having harmful effects in other countries too. In 1984 I visited both the Republic of Ireland and Greece. In both countries I saw the baleful effects of CAP. In both it was paying farmers to reclaim some of the most beautiful and biologically rich habitats in Europe. Huge machines were to be seen draining wetlands and removing rocks from hillsides at the expense of the European taxpayer. The laudable aim was to help remote farming communities but the method used was unnecessarily destructive. The farmers could so easily have been paid the same amount to conserve landscape and wildlife as well as to produce food. By pursuing our own national interests through reform of the CAP we will also be helping our neighbours to conserve their heritage too.

There are even more important things to do in the short term: it is crucial that we get our priorities right in the world as a whole. Therefore we must ask two fundamental questions. Where is the greatest concentration of life? Which habitats are under greatest threat? The first can be answered with certainty. The tropical rain forests hold far the greatest number of species; of all the habitats which mankind has inherited from the past, these forests hold the greatest potential. They are also under far greater threat than temperate forest, tundra, savanna or desert. It has been estimated that they are being destroyed at the rate of 12 ha (30 acres) a minute, that is an area the size of all of Britain's National Nature Reserves every 9 days. The reasons for this are well known. All the rain forests, except those of Australia, occur in Third World countries. In almost every nation there has been a great population increase. In most, the extra mouths could be fed by increasing productivity on the existing farmland; in most cases this has not, however, been achieved. More forest has therefore been cut down to provide land for unproductive and wasteful forms of agriculture. Also, the demand for tropical hard-woods by the developed nations could provide nations with rain forest a valuable market indefinitely, if measures are taken to renew the resource. Sadly this rarely happens. Nation after nation sells its birthright for a mess of potage. The get-rich-quick contractor is given a concession, he takes what he wants in the cheapest way possible and leaves devastation behind him – at best mutilated secondary forest which will take many years to recover, at worst a bare hillside which has lost all its soil from erosion and which can never grow trees again (Fig. 46).

Fig. 46. Foothills of the Himalayas, India. These hills were once covered by forest which could have provided timber crops inde-finitely and so have produced wealth and employment and as well would have conserved thousands of species of wildlife. Through unsuitable agricultural development they have lost their soils and have become a desert which cannot now be farmed and which is of little value for wildlife.

I have seen these things happen in every tropical country I
have visited. It was particularly sad to see it in Madagascar. This
great island became separated from the ancient southern continent of Gondwanaland 100 million years ago. As it drifted
further from what is now southern Africa and India it became
increasingly isolated and as a result its plants and animals
evolved separately from the rest of the world, and today most of
them are endemics: they are found in Madagascar and nowhere
else. The lemurs are the best-known examples (Fig. 47). Until
man arrived there from Indonesia about 1500 years ago,
Madagascar was covered by forest. In consequence, virtually all
the Madagascan plants and animals depend on forest for their
survival. Today only 20% of the forest remains and it is rapidly

Fig. 47. Maki (*Lemur catta*). Lemurs, like most of the plants and
animals of Madagascar, are only found there and nowhere else. The
diminishing forests on which they depend could be saved if farming
could be made more productive elsewhere on the island.

disappearing before slash and burn cultivation. The forests of Madagascar have already contributed the glorious Flame Tree which gives colour to so many towns in the tropics, and the Madagascar periwinkle which contains a drug which is important in the treatment of cancer. Madagascar cannot afford to protect its unique heritage unaided. If the biological wealth of this marvellous country is to be held in trust for mankind, mankind must help Madagascar retain its remaining forests.

Australia is another huge island, whose separation from the rest of the world has led to its developing an 'alternative' flora and fauna consisting largely of endemic species. Only a small part of eastern Australia is wet enough to allow rain forest to grow there. Like the rain forest of Madagascar it is unique and full of species of known and potential value. Yet huge areas of lowland rain forest have already been cleared and the remnants are under threat. When I visited Australia in 1972 I was appalled that a rich nation like Australia, which had no population problem, could not protect its rain forest. Since then, its conservation movement has grown considerably and at the eleventh hour there is now hope that some at least of the Australian rain forest will be conserved. If Australia, which is a very rich country, nearly lost its rain forest, the Third World Countries which are very poor cannot be expected to conserve theirs without help from the world community. How can we in Britain help about this and other urgent international conservation problems?

First we should strengthen the existing channels for international cooperation which already exist. Much thought has already gone into the production of the World Conservation Strategy (WCS). This was drawn up by the International Union for the Conservation of Nature and Natural Resources (IUCN), the World Wildlife Fund and the United Nations Environmental Program. It carefully related conservation with development, fully recognising that, in the long run, these support each other and do not conflict. So far the British government's response and support of the WCS has been meagre and disappointing. Here is a chance, perhaps the last chance to achieve effective conservation on a world basis. A government that fails to take it will be condemned by future generations, who will have to suffer our generation's neglect of one of the most important issues of its time.

Britain has special responsibilities, and should often take the lead in world conservation matters. For example, our involvement with Antarctica and the Falkland Islands brings special opportunities and duties in conserving the vast oceanic ecosystem of the Southern Seas which is based on prawn-like creatures known as krill. By our ensuring that Antarctica remains international and the krill is not overexploited this vast resource of protein can be managed indefinitely for mankind. By historical accident Britain is still in charge of numerous small islands with endemic flora and fauna and important colonies of sea birds, turtles and seals. We can work for their survival. The most enduring links of the Commonwealth are those of scientific and technological cooperation. In our own relatively impoverished islands we have developed valuable conservation technology. We should share this with our fellow Commonwealth countries on a much greater scale than hitherto.

International organisations are not necessary for much international conservation. We can do a great deal by developing the ecological aspects of our trade practices with other countries. We could make it clear to all who sold us timber from rain forest that we wished to continue this trade in the years ahead, and so we would only buy timber from firms which guaranteed to manage their forests as renewable resources. In the short term this would cost more, but when others had exhausted their supplies, the advantages of a more prudent policy would become obvious.

We should give much more attention to the possible effects of our exports of pesticides. It is clearly reprehensible to sell to other countries chemicals which we have banned on safety grounds in this country. There are exceptional cases where pesticides which are banned in Europe should be used in the tropics but, in general we should not sell those which we do not use ourselves.

It is obvious that there is much Britain can do to conserve its wildlife and that of other countries in the short term. In a previous chapter I have emphasised that it is important to put our own house in order because we shall not be listened to if we do not. Also, in the process of conserving wildlife at home we learn techniques which can be used overseas. Therefore the short-term strategy of protecting both sites and habitats in the wider countryside in Britain helps the short-term strategy overseas.

On its own, each element of the short-term strategy seems relatively small, whether it is site protection, the modification of CAP, the modification of the machinery of government, greater support for the World Conservation Strategy, increased overseas advisory work or modifications of trading practice; together, however, these elements make up a considerable package of endeavour. It is most unlikely that private citizens, firms or government bodies acting separately can generate enough steam to get it launched. Certainly market forces are ill fitted for achieving long-term conservation goals because they nearly always favour short-term solutions. There should be no disagreement between the political left and the political right that government must take the lead over conservation, as it does already with other activities concerned with the future, notably defence and preventive medicine. With these points in mind there appears to be a pressing need for an explicit statement of national intent. A clear statement should be made that conservation will henceforth be a national objective. The lack of a written constitution makes this more difficult than it otherwise might be. Each government in Britain outlines its plans in the Queen's speech at the opening of Parliament. The Queen's speech appears to be the best available vehicle for such a statement. Every subsequent Parliament should reaffirm the nation's commitment to conservation outlining the government's specific measures to support it. In this way conservation would be kept before the public's eye and would be linked with other national objectives.

Inevitably, future needs will compete with present needs to some extent. Increasingly each government will have to give more thought to getting the balance right. In doing this it will have to bear in mind the inherent bias against the long term which results from market forces tending to favour cheap, short-term solutions. As electorates become more sophisticated, governments which fail to give adequate support to future requirements, whether they concern conservation or the development of new industries and technologies, will suffer at the polls. When that happens we can be sure that the long-term strategy of conservation is beginning to work and the short-term one has held the fort to some purpose.

23

*

Education for the future

The long-term future depends on a change of attitude about conservation. Conservation cannot be achieved by elitist groups, however powerful. A common motivation must be developed and shared. We must try to bring about a change of attitude in everyone, but obviously the people who matter most are the young: everything depends on them. Therefore the main effort to change attitudes must come through the educational system.

Before discussing what might be done at the different levels of teaching, we should consider the psychological base of conservation. The case for effective conservation of our living resources is overwhelming, and is what commonsense demands if mankind considers himself within the context of time. However, introspection and observation of my fellow conservationists suggests that our strongest motivation is often something else. We may dislike the whimsical overtones of the term 'nature lovers', but that is exactly what we are: we love nature and so we want it to prosper. In other words, the aesthetic motivation is extremely important, and probably the most effective driving force behind the practice of all forms of conservation. Conservationists should not be ashamed of this; the objective of conserving wildlife because it is beautiful, and we want ourselves and future generations to be able to enjoy it, is admirable. In no way does it conflict with the need to conserve wildlife because our future physical well being also depends upon it. However, the aesthetic motivation is not enough on its own. There are dangers in overemphasising it, because there is a tendency for cynics to write off conservation if it is considered to be only a matter of aesthetics, and therefore something of marginal importance to most people today. Conservationists should be more aware of this danger. Nevertheless I believe that we should look at aesthetic motivation positively, and learn from the common experience of conservationists.

Those teaching conservation, far from decrying aesthetic motivation, should build on the inherent delight that every young child seems to have for the natural world. Once care for nature has been established in the mind of the child, he or she will more readily comprehend the other objectives of conservation, and ultimately will understand that conservation is the proper response to more than one human need. Educationalists should use the love of nature as the base from which to build a more realistic understanding of nature and man's position within it.

The overall educational aim is simple enough. The whole system from primary to tertiary levels and beyond should ensure that all are made aware of man's place in the natural world and his special responsibilities at this particular time. Enough factual knowledge should be acquired so that the requirements of living organisms are broadly understood. But far more important are attitudes, and particularly attitudes about the relationship between conservation, economics, politics and the arts. Huge barriers still exist between 'the two cultures'. Many highly intelligent men and women do not understand the relationship between the two. Few would disagree with Alexander Pope that 'the proper study of mankind is man', but many do not perceive that man cannot be understood without reference to his place in the biological scheme of things and to his environment. Biology must not be thought of as a subject fit for specialists alone. Everyone should be taught enough to see its relevance to all our activities. This may seem platitudinous to some, but as more and more people live in cities the less obvious are the links which bind us to other life on the planet. We must all learn to live in the real world.

It is easy to generalise, harder to particularise the details of what should be done. I cannot use my own experience as a pupil to support any particular approach to teaching since, as mentioned earlier, I received no formal education in biology until I went to university. However, my experience does give a great deal of support for the idea that attitudes matter enormously. Children are no fools: they can sense enthusiasm, disparagement and scorn and are greatly influenced by what they sense. No amount of formal teaching of biology and conservation in schools will do any good if those who teach it do not believe in what they are saying. Fortunately an increasing proportion of our teachers were made enthusiastic about environmental matters when they were students in the 1960s and 1970s. They are especially suitable people to make children enthusiastic about conservation today.

At the level of primary education, the job seems relatively easy. Most children start by being curious and they delight in the extraordinariness of the natural world. They do not need much encouragement so long as they can actually see, hear and (above all) touch plants and animals. The provision of ponds and other natural habitats on land attached to the school, and the opportunity to visit the country whether in the form of educational nature reserves or ordinary farms and woods is very important. Far too little is done by most schools, even by those in the country, to use these resources. At this stage, simple contact with nature is what matters most.

The real problem comes with secondary education. A combination of having to face the problems of adult life and adolescent revolt frequently destroys what has been built up in the primary school days. At Monks Wood Experimental Station we frequently had Open Days for schools and the general public. It was always a delight to show the younger children round – they were really interested and asked the most fundamental questions – questions which neither we nor anyone else could have answered adequately. With rare exceptions the older children were completely different. Even if interested, they were determined to appear bored and supercilious. One of the reasons was all too obvious – the bored master or mistress accompanying them. The subject matter of our exhibits did not quite fit their curricula; their main concern was to avoid being caught out in front of their pupils.

The difficulties of changing over from the freedom of primary education to the rigours of secondary education are well known to the teaching profession. They are more likely to be overcome by changes in the attitude of teachers rather than by changes in curricula.

However, changes in curricula are also needed. In the past, classical biology based on the study of comparative anatomy and physiology within an evolutionary framework could be used as a base for teaching conservation, although it rarely was. Today the emphasis on cellular mechanism makes this more difficult, so that even students taking biology for their 'O' and 'A' levels can go through courses without being faced with conservation problems. Even if this were to be remedied by altering the content of the courses, it would only result in those specialising in biology learning about conservation. What is needed is that everyone should learn about it, and particularly those who will wield power in politics, administration and business. It should not be too difficult to provide a course which teaches enough about evolution, ecology and conservation to show their relevance to everyone. This should be an aim of all secondary education.

For some years, conservation has been taught at some universities in the United Kingdom. It is essentially a postgraduate subject provided for those who hope to get a job in conservation work. The MSc course run by University College, London, for example, has been outstandingly successful in producing student generation after student generation of young men and women who have got to grips with the conservation problems of their time. There are a fair number of environmental courses but, by and large, universities are not good at integrating environmental topics with other subjects, even where one would expect the relevance of environment and conservation to be obvious as in biology and agriculture courses. I fear this may be due to conservation being both an applied science and popular among non-academics. Certainly conservation and ecology attract a certain amount of pseudoscience, so it is not altogether surprising that more timorous academics retire into their safe fields of theoretical studies and shun contact with environmentalists. Again, I believe that a general change of attitude is vastly more important than a change of curricula. Historians, lawyers, economists, as well as foresters and agriculturalists, should learn the relevance of conservation to their subject. If conservation ideas were inserted into all these disciplines it would open up exciting new dialogues within universities and make students far more receptive to conservation requirements, when they themselves became influential in government, industry and in the educational world.

Two types of educational establishment are particularly important – the Colleges of Education which train the teachers and the Agricultural Colleges which train farmers and land managers. In both the need is less to initiate special courses in conservation than to ensure that conservation is introduced into the many courses where it is relevant. I know from personal experience how difficult this is to achieve in practice. Those who run the courses into which conservation should be injected claim, often with much justice, that they are not competent to teach conservation and that their courses are too full anyway. They ask what they should drop to make room for conservation. Even at the best agricultural colleges, conservation tends to be thought of as an extra which can be dealt with in a supernumerary lecture rather than something which should be introduced into lectures on drainage, fertilisers, pesticides and land management. Parallel courses on conservation are often run, but they do not fill the bill as the students who attend them are those who aim to go into environmental not farming jobs. These problems are well appreciated now, not least by FWAG. The Conservation in Agricultural Education Guidance Group working under the aegis of FWAG has produced leaflets on the conservation aspects of each farming topic; these are being used increasingly by lecturers in agriculture.

To conclude, there are many means by which environmental education for all can be improved, but the prime need is for a change of attitude within the whole educational system. This can only be achieved by those pioneers already in the system who are truly convinced that all education is seriously deficient if it does not help the next generation to face the great environmental challenges of our age.

24

Long-term conservation strategy

Conservation will get more difficult. It will not be achieved unless it is supported by the majority, and it will not be supported by the majority unless its importance is understood and accepted. In carrying out 'the minute particulars' of the holding operation, which were outlined in Chapter 22, conservationists must not lose their vision of the objective to which they are committed. If they allow it to become obscured they will not be able to bring about that change of attitude on which the future of conservation depends.

We should never forget that the objective of conservation is no less than to maintain the living resources of the world so that each generation can use and enjoy them. Conservationists must make it quite clear that in pursuing this objective they seek a new dimension in public life. Conservation should come to be accepted as a matter of course in the way that we already accept the need for peaceful relations between nations, improved safety standards at work and improved medical care. Once this is achieved the future of our living resources will be assured.

The grand objective must be firmly linked with practicalities so that it becomes manageable. We can build on those features of the future which we can predict. We have already seen that whatever way the world develops, conservation will always be effected by the two types of activity which characterise it today. We shall always have to set aside particular areas in which the conservation of wild plants and animals is the primary land use: such places will always be necessary whether we call them nature reserves, national parks, wildlife areas or whatever.

Secondly, we shall always have to integrate conservation with other kinds of land use on the land outside the nature reserves.

The potential for conflict between conservation and other activities will remain because there is bound to be competition for land between different land users. Likewise, the degree to which conservation should be allowed to modify agricultural, forestry and industrial activities will always be a matter for debate, and the outcome will always be a matter of compromise between partially incompatible objectives. However, conflicts of interest can be reduced greatly by adopting sensible codes of practice, so long as these are backed by public opinion; there has to be a genuine desire to achieve the best result in each situation.

Conflicts can only be resolved effectively when the relevant facts are known. Even today many have been reduced or solved by the preparation of Environmental Impact Assessments. These describe the resource at risk and assess the impact of the proposed development upon it. In the future, the environment should always be considered when major new developments are proposed, and Environmental Impact Assessments made. The procedure should not be confined to local proposals such as siting power stations, building tidal barrages, and reclaiming coastal land, but should also be used when changes in agricultural and forestry policy are proposed and before new techniques are introduced on a large scale. A public convinced of the importance of conservation would insist on Environmental Impact Assessments being made whenever it was needed, and on a fair apportionment of the costs of remedial action between developer and taxpayer when decisions have been reached.

Whatever system is used to integrate conservation with other land uses and activities, it will result in both successes and failures. It will be necessary to learn from them, and therefore monitoring will be an essential component of future conservation activity. Changes in the area and quality of habitats will have to be recorded, as well as fluctuations in the populations of species which are especially valued and of those which can be used as indicators of environmental health. Publicity should be given to the results of monitoring programmes. This will help to maintain interest in conservation once it has become a normal part of ordinary life. People will want to follow the environmental state of play in the way that today they follow the latest trade figures, financial indices and the fortunes and form of football clubs and athletes. Already the annual reports and progress reports of the Nature Conservancy Council, the Countryside Commissions, the Royal Society for the Protection of Birds and the Nature Conservation Trusts provide great satisfaction and encouragement to their readers and point the way ahead.

Publication of the results of monitoring will do much to keep conservation in the public eye, but quite explicit statements about conservation will also be necessary. The speech from the throne, party political manifestos, and annual reviews by companies and leaders in agriculture, forestry and commerce will provide opportunities.

I am confident that today's conservation movement can succeed in being the catalyst for developing conservation mindedness in the population as a whole. Success is not a foregone conclusion and it will depend at least partly on a change in the attitude of conservationists. Today, conservation gets very small slices of the national and international cakes. It is rarely put into perspective; the resources devoted to it are derisory when measured against what is urgently needed. As a result, those who understand the relevance of conservation and its importance have to shout in order to be heard. They have learnt that results are only achieved when matters are brought to a head by presenting issues crudely, even contentiously. I believe that the adversarial approach has been necessary in the early days of the conservation movement, but it will not be good enough in the future. It is always easier to unite against something than for something. In the case of conservation this has led to a polarisation of attitudes which could be a serious obstacle in getting it accepted widely.

There will always be a place for prophetic gadflies which sting complacent society into action, even if they do oversimplify and exaggerate. On the other hand, the conservation movement as a whole cannot hope to share its aspirations with the rest of the world if its main characteristic appears to be 'anti'. Conservationists must put more emphasis on the positive. Some will find it very difficult to forsake the safety provided by an ethos of shared opposition. A more positive approach demands a tougher skin as well as more imagination, and entails taking on more responsibility.

The attitude which is now necessary is one which includes sympathy for the aims of conservation as well as for the interests with which conservation competes. I believe that the development of FWAG is highly significant because it shows that such an attitude is attainable. FWAG could become just another conservation body, or just a conscience-salving front for the agricultural industry, but that is not the intention. FWAG is aiming at real partnership between conservation and agriculture. So when we appoint its county advisers we look out for young men and women who have sympathy both with farmers and with conservation. We start with the assumption that both farming and conservation matter and practical ways must be found to relate them so that both can be achieved on the farm. Experience has shown that enthusiasm for conservation need not be lost when we drop an adversarial position and devote our energies to finding practical ways of achieving effective conservation within productive farming systems.

At first, taking conservation seriously may appear to be a dream of Utopia, but when the requirements are broken down into manageable portions it becomes an entirely practicable objective. Conservationists should set their sights high; I believe they may be surprised how quickly they succeed.

25

---
*

The Bird is on the Wing . . .

The experience of my working life in conservation has led me to believe that we should give more thought to time and to our generations' position in it. If we take time into account we cannot escape the conclusion that conservation matters and that our generations have a unique responsibility to undertake certain specific actions in its support. This book is a plea to take conservation seriously.

We should not delude ourselves: we do have a genuine choice and it is time to act. We can predict in general terms what will happen if we do not face up to our responsibilities. Initially all will appear to be reasonably well, because the environmental movement has already got up enough steam to produce numerous local successes; new nature reserves will be established, pollution will be reduced locally and improvements will be made in areas such as in the trade of wild plants and animals. However, if nature conservation continues to be regarded merely as an extra, a legitimate but peripheral activity which must be catered for in a civilised state as are the fine arts and opera, it will not be able to deal with the major problems which will emerge. For, unless conservation is supported by the majority as a matter of principle and long-term commonsense, it will suffer a series of defeats when confronted by pressing economic needs. Each defeat will seem rather insignificant but in total the effect will be very serious. Failure will lead to worse failure and remedial action will become increasingly difficult and expensive. Expectations will be lowered and it will become harder and harder to get out of this vicious spiral.

We cannot predict the biological details of the events which will occur or their exact timing but their general pattern is inevitable if we do not face up to our responsibilities now. There will be a gradual erosion of our living resources until the extinction of a species is so commonplace that it will no longer be news. As things get worse, major catastrophes will become imminent. These may include major climatic shifts, the poisoning of the seas due to unwise developments on land or on the sea bed itself and the rapid spread of diseases in species on which man directly depends. Once major threats are identified and are obvious for all to see, then some action will be taken. It may be taken just in time to avert catastrophe but procrastination will mean that much will have been lost unnecessarily, and remedial action will be much more expensive than it need have been. On the other hand, remedial action may not be taken in time. It will be too late to prevent one major catastrophe from triggering off others. As life on the planet becomes increasingly impoverished, even the resources for day-to-day survival will become reduced. There will be international recrimination and a grab for what is left. Wars will seem the less terrible option.

I think that this almost unthinkable scenario must be taken seriously, but I do not believe it is inevitable. We really do have the choice and the opportunity to change man's perception of the world he lives in. A combination of stick and carrot is always more effective than either alone. Much has been made of the stick of fear, but nothing like enough of the carrot provided by a positive, widely held philosophy of conservation.

Conservation of the world's living resources could provide the common motivation for international cooperation which is so sadly lacking today. To some extent peace and improved standards of living can be obtained unilaterally by single nations, but conservation can only be achieved by international cooperation. The whole of mankind shares the wish that future generations should prosper. We all want to do what we can for our grandchildren. Unlike some of our neighbours today future generations pose no threat to ourselves.

If we need the stimulus of a common enemy against whom we can unite, we have one – the callous spirit of individuals and organisations which ignores the needs of the unborn.

A wise use of land was once an integral part of numerous

cultures throughout the world. Even in the developed world, where technology has yet to learn how to support the future as well as the present, numerous foresters and farmers still yearn to do both. The old love and care for land is latent but still exists.

A positive philosophy of conservation can be supported by all mankind; it accords particularly well with the Christian belief in stewardship and the Hindu's reverence for life. Conservation is a glorious exception in an age enfeebled by cynicism and pessimism, because it courageously asserts a faith in the future. If we all become conservationists and accept that the earth is our heritage, we shall look on future generations as our neighbours. The possibilities which will result from our loving them are tremendous.

POSTSCRIPT

---- * ----

This book was written at the end of a long period in which the management of the British countryside changed relatively little. Powerful governmental support for agriculture dominated the scene and led to high land prices. The water and forestry industries were largely government controlled. The operation of the Pesticides Safety Precautions Scheme had remained unchanged for decades. Governmental support for conservation continued at a very low level.

In recent months the log jam of economic and political constraints has been broken. The public has become impatient with the butter and cereal mountains and has become aware that disproportionate support for agriculture is causing serious damage to the environment. All this has led to a radical reappraisal of the Common Agricultural Policy, and important changes in organisation have been made or proposed. In addition to these matters, which will obviously affect conservation directly, most people realise that the old patterns of economic life and employment are breaking down and are being replaced by new ones based on new technologies. In these circumstances how relevant is the experience of the recent past?

The future looks more flexible: will the problems of maintaining genetic diversity and of pollution control become easier to solve in the future or will they remain, but in a different guise? I suspect the latter. If so, the effectiveness of conservation in the future will still depend on political will which, in turn, will depend on the perceptions and attitudes of people rather than on changes in circumstance and organisation. For example, the recently passed legislation on the control of pesticides (Food and Environment Protection Act, 1985) will only be effective if enough inspectors are appointed, but they will not be appointed unless the agricultural departments really intend to reduce the current amount of misuse of pesticides.

There are proposals to privatise the water industry. However, the control of pollution cannot be achieved through the play of market forces; new statutory arrangements will have to be made. Whether they will be better or worse than existing ones will depend on government's and industry's perception of public concern. It will be much harder to control pollution if the water industry is privatised, but almost any organisational ploy will succeed if the will exists.

It is too soon to predict the effects of diminishing governmental support for agriculture. Reduction in grants for drainage is bound to help conservation, on the other hand a poorer farmer has less money to spend on conservation. There has been a welcome increase in the annual grant of the Nature Conservancy Council (NCC) from £22.7 million in 1985–6 to £32.1 million in 1986–7; this is already enabling the NCC to give better protection to SSSI by means of management agreements. The Forestry Commission's new policy for broad-leaved woodlands, and above all its explicit recognition of the value of ancient woodland, has led to their providing an enlightened Broadleaved Woodland Grant Scheme. But the extent to which land taken out of cultivation can be afforested will depend on the support government can give farmers in the years before timber can be sold.

The first significant sign that conservation is to be integrated with agriculture is apparent in the Agriculture Bill now before Parliament. Not only must agricultural departments 'endeavour to achieve a reasonable balance' between agriculture and conservation, but they can give grants to farmers to manage their land for conservation in designated 'Environmentally Sensitive Areas' (ESA). This could lead to much more harmonious agricultural practices, but whether this enlightened provision will have a significant effect will depend on the number of ESA designated; this will depend on the extent to which government takes conservation seriously.

Greater variety and flexibility in land use is encouraging

conservationists to become more interested in re-creating habitats as well as conserving those which cannot be re-created. This is bound to lead to greater cooperation with other users of land, who will become increasingly involved with conservation.

The scope for non-agricultural uses of farmland will increase as land prices fall but, unless planning controls are maintained and in some respects strengthened, piecemeal exploitation of woods and disused farmland could be harmful to conservation and landscape alike. On the other hand, cheaper land will give opportunities to conservation bodies to acquire more nature reserves. Whether they can seize these opportunities will depend on the financial support they get from individuals, industry and government. It is encouraging that all three of the main political parties are now vying with each other to get the 'green' vote.

Despite all these changes, the basic problems remain. Pollution cannot be controlled without cost. The selection of National Nature Reserves and SSSI will remain a necessary and complicated process of assessing biological need in changing circumstances. Conservation will continue to compete with other interests for land, even if competition with agriculture is reduced. Recent developments have altered the political landscape and provide new opportunities, but they do not alter the main conclusion of past experience, which is that conservation must be recognised as a major national and international objective, and must be integrated with other objectives if it is to succeed. This will still depend upon public perception and will.

In recent months the world has felt a more dangerous place, because the relations between the superpowers have deteriorated and because of the Chernobyl disaster in the Ukraine. The disaster has produced doubts about all nuclear reactors and has convinced most people that the world could not survive a nuclear war. These events reduce faith in the future and hence in the value of conservation. Yet they could be used constructively. If we had the foresight and courage to put conservation of all our resources at the top of the agenda we would reduce world tension and make economic and defence problems easier to solve. Omar Khayyam's 'Bird of Time' urged us to take opportunities while they exist. The Bird is on the wing . . .

REFERENCES

--------------------------------- ★ ---------------------------------

A complete list of references is not practicable in a book of this kind. The list provided here includes
(a) books for general reading on the subjects covered
(b) journals which are particularly relevant to the subject matter
(c) scientific papers supporting particular statements in the book.
Those in categories (a) and (b) are shown with an asterisk.
The references are numbered to enable the identification of those relevant to each chapter, as detailed below.

Chapter

1	12, 33, 44, 47.
2	45, 55, 56, 67, 78, 81, 123, 125, 126, 134, 138, 143, 144, 150, 155.
3	13, 15, 39, 40, 42, 49, 53, 71, 85, 89, 106, 110, 115, 123, 136, 146, 164.
4	11, 24, 26, 51, 54, 59, 72, 77, 91, 98, 99, 124, 125, 129, 134, 139, 163
5	27, 77, 85, 97, 107, 109, 111, 112, 113, 136, 167.
6	13, 39, 40, 51, 78, 85, 89, 106, 115, 136, 146.
7	13, 43, 51, 97, 98, 106, 113, 120, 121, 136, 141, 157.
8	10, 21, 77, 81.
9	28, 108, 167.
10	8, 46, 66, 76, 87, 88, 145, 153, 154, 155, 156, 158.
11	2, 3, 46, 145, 156.
12	52, 79, 95, 105, 127.
13	7, 20, 35, 52, 60, 79, 90, 142, 147, 163, 169, 170, 171, 173.
14	14, 29, 30, 31, 32, 94, 100, 147, 151, 166.
15	23, 34, 36, 37, 38, 60, 63, 64, 65, 75, 79, 80, 83, 86, 90, 93, 94, 95, 101, 102, 103, 104, 105, 116, 128, 130, 132, 133, 135, 137, 140, 147, 152.
16	1, 4, 17, 20, 22, 23, 25, 29, 30, 31, 32, 33, 48, 73, 75, 86, 92, 93, 137, 147, 149, 159.
17	57, 68, 69, 80, 118, 119, 131, 147.

18	1, 17, 18, 61, 74, 82.
19	4, 22, 58, 62, 96, 107, 147, 165, 166.
20	9, 44, 77, 106, 161, 168.
21	19, 67, 81, 84, 121, 136, 155, 161, 167, 172.
22	16, 50, 70, 84, 114, 117, 126, 146, 148, 161, 162.
23	50, 77, 114, 117.
24	21, 50, 114, 117, 126, 161, 162.

1 Ackefors, H. (1971). Effects of particular pollutants, III. Mercury pollution in Sweden with special reference to conditions in the water habitat. *Proc. R. Soc. Lond.* B, **177**: 365–87.

2 Advisory Committee on Myxomatosis (1954). HMSO, London.

3 Advisory Committee on Myxomatosis (1955). *Myxomatosis*, second report. HMSO, London.

4 Advisory Committee on Pesticides and other Toxic Substances (1969). *Further review of certain persistent organochlorine pesticides used in Great Britain.* HMSO, London.

5 Advisory Committee on Poisonous Substances Used in Agriculture and Food Storage (1964). *Review of the persistent organochlorine pesticides. See* Cook, J. W.

6 Agriculture Act (1947). HMSO, London.

7 Alexander, W. B. (1932). The bird population on an Oxfordshire farm. *J. Anim. Ecol.*, **1**: 58–64.

8 Andersen, J. (1957). Studies in Danish Hare-populations. 1. Population fluctuations. Communication No. 21, Vildtbiologisk Station Kalø. *Dan. Rev. Game Biology*, **3**: 85–131.

*9 Ashby, E. (1978). *Reconciling Man with the Environment.* Oxford University Press, Oxford.

*10 Barber, D. (ed.) (1970). *Farming and Wildlife: a Study in Compromise.* Royal Society for the Protection of Birds, Sandy.

11 Beresford, J. E. and Wade, P. M. (1981). Field ponds in North Leicestershire: their characteristics, aquatic flora and decline. *Trans. Leicester Lit. Phil. Soc.*, **76**: 25–34.

12 Berry, R. J. (1983). Genetics and conservation. In *Conservation in Perspective*, eds A. Warren and F. B. Goldsmith, pp. 141–56. John Wiley & Sons, Chichester.

13 Bibby, C. J. and Tubbs, C. R. (1975). Status, habitats and conservation of the Dartford Warbler in England. *Brit. Birds*, **68**: 177–95.

14 Blackmore, D. K. (1963). The toxicity of some chlorinated hydrocarbon insecticides to British wild foxes (*Vulpes vulpes*). *J. Comp. Path. Ther.*, **73**: 391–409.

15 Blackwood, J. W. and Tubbs, C. R. (1970). A quantitative survey of chalk grassland in England. *Biol. Conserv.*, **3**: 1–5.

*16 Body, R. (1982). *Agriculture: the triumph and the shame.* Temple Smith, London.

17 Borg, K., Wanntorp, H., Erne, K. and Hanko, E. (1969). Alkyl mercury poisoning in terrestrial Swedish wildlife. *Viltrevy*, **6**: 301–76.

18 Bull, K. R., Murton, R. K., Osborn, D., Ward, P. and Cheng, L. (1977). High levels of cadmium in Atlantic seabirds and sea-skaters. *Nature, Lond.*, **269**: 507–9.
19 Cambridge Bird Club (1927–). *Reports 1927–*. Pembgate Ltd, Cambridge (previously Cambridge University Press and Severs, Cambridge).
*20 Carson, R. (1963) *Silent Spring*. Hamish Hamilton, London.
*21 Carter, E. (1983). The Farming and Wildlife Advisory Groups. *Biologist*, **30** (3): 124–6.
22 Cook, J. W. (Chairman) Advisory Committee on Poisonous Substances used in Agriculture and Food Storage (1964). *Review of the Persistent Organochlorine Pesticides*. HMSO, London.
23 Cooke, A. S. (1973a). Shell thinning in avian eggs by environmental pollutants. *Environ. Pollut.*, **4**: 85–152.
24 Cooke, A. S. (1973b). The effects of DDT, when used as a mosquito larvicide, on tadpoles of the frog *Rana temporaria*. *Environ. Pollut.* **5**: 259–73.
25 Cooke, A. S., Bell, A. A. and Haas, M. B. (1982). *Predatory birds, Pesticides and Pollution*. Institute of Terrestrial Ecology, Monks Wood Experimental Station, NERC, Cambridge.
26 Cooke, A. S. and Scorgie, H. R. A. (1983). *Focus on Nature Conservation*, No. 3. The status of the commoner amphibians and reptiles in Britain. Nature Conservancy Council, Shrewsbury.
27 Countryside Act (1968). HMSO, London.
28 Countryside Commission (1985). *Annual Report 1984–85*. Countryside Commission, Cheltenham.
29 Cramp, S. and Conder, P. J. (1961). *The Deaths of Birds and Mammals*. Royal Society for the Protection of Birds, London.
30 Cramp, S. and Conder, P. J. (1965). The fifth report of the Joint Committee of the British Trust for Ornithology and the Royal Society for the Protection of Birds on Toxic Chemicals. August 1963 to July 1964. Royal Society for the Protection of Birds, Sandy.
31 Cramp, S., Conder, P. J. and Ash, J. S. (1962). *Deaths of Birds and Mammals from Toxic Chemicals*. Royal Society for the Protection of Birds, London.
32 Cramp, S., Conder, P. J. and Ash, J. S. (1963). *Deaths of Birds and Mammals from Toxic Chemicals*. Royal Society for the Protection of Birds, Sandy.
*33 Darwin, C. (1859). *On the Origin of Species by Means of Natural Selection, or the Preservation of Favoured Races in the Struggle for Life*. Murray, London.
34 Davis, B. N. K., Moore, N. W., Walker, C. H. and Way, J. M. (1969). A study of millipedes in a grassland community using dieldrin as a tool for ecological research. In *The Soil Ecosystem*, ed. J. G. Sheals, pp. 217–28. The Systematics Association, London.
*35 Day, J. W. (1957). *Poison on the Land: the War on Wild Life, and Some Remedies*. Eyre & Spottiswoode, London.
36 Dempster, J. P. (1967). The control of *Pieris rapae* with DDT, I. The natural mortality of the young stages of *Pieris*. *J. Appl. Ecol.*, **4**: 485–500.

270   *References*

37 Dempster, J. P. (1968a). The control of *Pieris rapae* with DDT, II. Survival of the young stages of *Pieris* after spraying. *J. Appl. Ecol.*, 5: 451–62.
38 Dempster, J. P. (1968b). The sublethal effect of DDT on the rate of feeding by the ground-beetle *Harpalus rufipes*. *Entomol. Exp. Appl.*, 11: 51–4.
39 Department of Health for Scotland (1947). *National Parks and the Conservation of Nature in Scotland*. Report by the Scottish National Parks Committee and Scottish Wild Life Conservation Committee, Cmnd 7235. HMSO, Edinburgh.
40 Department of Health for Scotland (1949). Nature Reserves in Scotland, Final Report by the Scottish National Parks Committee and the Scottish Wildlife Conservation Committee. Cmnd 7814. HMSO, Edinburgh.
*41 Duffey, E. (ed.) (1968–). *Biological Conservation*. Applied Science Publishers Ltd, London.
*42 Duffey, E. (1974). *Nature Reserves and Wildlife*. Heinemann, London.
*43 Duffey, E., Morris, M. G., Sheail, J., Ward, L. K., Wells, D. A. and Wells, T. C. E. (1974). *Grassland Ecology and Wildlife Management*. Chapman & Hall, London.
*44 Ehrlich, P. and Ehrlich, A. (1981). *Extinction. The causes and consequences of the disappearance of species*. Random House, New York.
45 Ewen, A. H. and Prime, C. T. (1975) (Trans. Ed.). *Ray's Flora of Cambridgeshire* (Catalogus plantarum circa Cantabrigiam nascentium). Wheldon & Wesley, Hitchin.
*46 Fenner, F. and Ratcliffe, F. N. (1965). *Myxomatosis*. Cambridge University Press.
*47 Frankel, O. H. and Soule, M. E. (1981). *Conservation and Evolution*. Cambridge University Press.
48 George, J. L. and Frear, D. E. H. (1965). Pesticides in the Antarctic. *J. Appl. Ecol.*, 3 suppl.: 155–67.
49 Good, R. (1948). *A Geographical Handbook of the Dorset Flora*. Dorset Natural History and Archaeological Society, Dorchester.
*50 Green, B. (1981). *Countryside Conservation*. George Allen & Unwin, London.
51 Hammond, C. O. (Revised by Merritt, R.) (1983). *The Dragonflies of Great Britain and Ireland*, second edition. Harley Books, Colchester.
52 Hardy, A. R. and Stanley, P. I. (1984). The impact of the commercial agricultural use of organophosphorus and carbamate pesticides on British wildlife. In *Agriculture and the Environment*, pp. 72–80. Natural Environment Research Council, Cambridge.
53 Harrison, C. M. (1976). Heathland management in Surrey, England. *Biol. Conserv.*, 10: 211–20.
54 Hartke, W. (1951). Die Heckenlandschaft. Der geographische Charakter eines Landeskulturproblems. *Erdkunde*, 5: 132–52.
*55 Hawkes, J. G. (ed.) (1978). *Conservation and Agriculture*. Duckworth, London.
*56 Hawksworth, D. L. (ed.) (1974). *The Changing Flora and Fauna of Britain*. Academic Press, London.

57 Holdgate, M. W. (ed.) (1971). The sea bird wreck in the Irish Sea, autumn 1969. *Publication Series C*, no. 4, The Natural Environment Research Council, London.
58 Holdgate, M. W., Kassas, M. and White, G. F. (eds) (1982). *The World Environment 1972–1982*. A report by the United Nations Environment Programme. Tycooly International Publishing Ltd, Dublin.
59 Hooper, M. D. (1970). Dating hedges. *Area* (4): 63–5.
60 Hunt, E. G. and Bischoff, A. I. (1960). Inimical effects on wildlife of periodic DDD applications to Clear Lake, *Calif. Fish Game*, **46**, 91–106.
61 Hutton, M. (1982). *Cadmium in the European Community: a Prospective Assessment of Sources, Human Exposure and Environment Impact*. Monitoring and Assessment Research Centre, Report 26, London.
62 International Union for the Conservation of Nature and Natural Resources (1970). *IUCN, Eleventh Technical Meeting Papers and Proceedings*. New Delhi, India. 25–28 November 1969, Vol. 1. IUCN Morges, Switzerland.
63 Jefferies, D. J. (1973). The effects of organochlorine insecticides and their metabolites on breeding birds. *J. Reprod. Fert.*, Suppl. 19: 337–52.
64 Jefferies, D. J. (1975). The role of the thyroid in the production of sublethal effects by organochlorine insecticides and polychlorinated biphenyls. In *Organochlorine Insecticides: Persistent Organic Pollutants*, ed. F. Moriarty, pp. 132–230. Academic Press, London.
65 Jefferies, D. J. and Davis, B. N. K. (1968). Dynamics of dieldrin in soil, earthworms and Song Thrushes. *J. Wildl. Mgmt*, **32**: 441–56.
66 Jefferies, D. J. and Pendlebury, J. B. (1969). Population fluctuations of stoats, weasels and hedgehogs in recent years. *J. Zool. Lond.*, **156**: 513–17.
*67 Jenkins, D. (ed.) (1984). *Agriculture and the Environment*. Natural Environment Research Council, Cambridge.
68 Jensen, S. (1966). Report of a new chemical hazard. *New Scient.*, **32**, 612.
69 Jensen, S., Kiahlström, J. E., Olsson, M., Lundberg, C. and Örberg, J. (1977). Effects of PCB and DDT on Mink (*Mustela vison*) during the reproductive season. *Ambio*, **6**: 239.
*70 Jolly, A. (1980). *A World Like Our Own. Man and Nature in Madagascar*. Yale University Press, New Haven and London.
71 Jones, C. A. (1973). *The Conservation of Chalk Downland in Dorset*. Dorset County Planning Department, Dorchester.
72 Jones, R. C. (1971). A survey of the flora, physical characteristics and distribution of field ponds in north east Leicestershire. *Trans. Leicester Lit. Phil. Soc.*, **65**: 12–31.
73 Keynes, G. (ed.) (1983). *Blake, Complete Writings*. Oxford University Press, Oxford.
74 Kobayashi, J. (1971). Relation between the 'Itai-Itai' disease and the pollution of river water by cadmium from a mine. *Fifth Intl Water Poll. Res. Conf.*, San Francisco. July 1970. Pergamon Press, New York.
75 Lockie, J. D., Ratcliffe, D. A. and Balharry, R. (1969). Breeding success

272   *References*

and organochlorine residues in Golden Eagles in West Scotland. *J. Appl. Ecol.*, **6**: 381–9.
76 Lockley, R. M. (1954). The European rabbit-flea *Spilopsyllus cuniculi*, as a vector of myxomatosis in Britain. *Vet. Rec.*, **66**: 434–5.
*77 Mabey, R. (1980). *The Common Ground*. Hutchinson, London.
*78 Macarthur, R. H. and Wilson, E. O. (1967). *The Theory of Island Biogeography*. Princeton University Press, Princeton.
*79 Mellanby, K. (1967). *Pesticides and Pollution*. Collins, London.
*80 Mellanby, K. (ed.) (1970–). *Environmental Pollution*. Applied Science Publishers Ltd, London.
*81 Mellanby, K. (1981). *Farming and Wildlife*. Collins, London.
82 Metal Bulletin Ltd (1978). *Cadmium 77. Proceedings, First International Cadmium Conference, San Francisco*. January 1977. Metal Bulletin Ltd, Worcester Park, Surrey.
83 Milstein, P. le S., Prestt, I. and Bell, A. A. (1970). The breeding of the Grey Heron. *Ardea*, **58**: 171–257.
84 Ministry of Agriculture, Fisheries and Food (1975). *Food from our own Resources*. Cmnd 6020. HMSO, London.
*85 Ministry of Town and Country Planning (1947). *Conservation of Nature in England and Wales*: Report of the Wild Life Conservation Special Committee (England and Wales). Cmnd 7122. HMSO, London.
*86 Monks Wood Experimental Station (Reports for: 1960–65; 1966–68; 1969–71; 1972–73; Supplementary Report for the year 1966). The Nature Conservancy, NERC, London.
87 Moore, N. W. (1956). Rabbits, buzzards and hares: two studies on the indirect effects of myxomatosis. *Terre vie*, **103**: 220–5.
88 Moore, N. W. (1957). The past and present status of the buzzard in the British Isles. *Brit. Birds*, **50**: 173–97.
89 Moore, N. W. (1962a). The heaths of Dorset and their conservation. *J. Ecol.*, **50**: 369–91.
90 Moore, N. W. (1962b). Toxic chemicals and birds: the ecological background to conservation problems. *Brit. Birds*, **55**: 428–35.
91 Moore, N. W. (1962c). Our disappearing hedges. *Countryside*, **19**: 321.
92 Moore, N. W. (1964). Intra- and interspecific competition among dragonflies (Odonata): an account of observations and field experiments on population density control in Dorset, 1954–60. *J. Anim. Ecol.*, **33**: 49–71.
93 Moore, N. W. (1965). Environmental contamination by pesticides. In *Ecology and the Industrial Society*, ed. G. T. Goodman, pp. 219–37. British Ecological Society, Symposium 5. Blackwell, Oxford.
94 Moore, N. W. (ed.) (1966). Pesticides in the environment and their effects on wildlife. The Proceedings of an Advanced Study Institute sponsored by the North Atlantic Treaty Organization. Monks Wood Experimental Station, England. 1–14 July 1965., *J. Appl. Ecol.* vol. 3, Supplement. Blackwell Scientific Publications, Oxford.
*95 Moore, N. W. (1967). A synopsis of the pesticide problem. *Adv. Ecol. Res.*, **4**: 75–129.

96 Moore, N. W. (1970). Pesticides know no frontiers. *New Scient.*, **46**: 114–15.

*97 Moore, N. W. (1982). What parts of Britain's countryside must be conserved? *New Scient.* **93**: 147–9.

98 Moore, N. W. (1986). Acid water dragonflies in eastern England – their decline, isolation and conservation. *Odonatologica* (in press).

99 Moore, N. W., Hooper, M. D. and Davis, B. N. K. (1967). Hedges I. Introduction and reconnaissance studies. *J. appl. Ecol.*, **4**: 201–20.

100 Moore, N. W. and Ratcliffe, D. A. (1962). Chlorinated residues in the egg of a Peregrine Falcon (*Falco peregrinus*) from Perthshire. *Bird Study*, **9**: 242–4.

101 Moore, N. W. and Tatton, J.'Og. (1965). Organochlorine insecticide residues in the eggs of sea birds. *Nature, Lond*, **207**: 42–3.

102 Moore, N. W. and Walker, C. H. (1964). Organic chlorine insecticide residues in wild birds. *Nature, Lond.*, **201**: 1072–3.

103 Moriarty, F. (1968). The toxicity and sublethal effects of p.p'-DDT and dieldrin to *Aglais urticae* (L.) (Lepidoptera: Nymphalidae) and *Chorthippus brunneus* (Thunberg) (Saltatoria: Acrididae). *Ann. appl. Biol.*, **62**: 371–93.

104 Moriarty, F. (1969). The sublethal effects of synthetic insecticides on insects. *Biol. Rev.*, **44**: 321–57.

*105 Moriarty, F. (1975). *Pollutants and Animals: A Factual Perspective*. George Allen and Unwin, London.

106 National Parks and Access to the Countryside Act (1949). HMSO, London.

107 Nature Conservancy Council Act (1973). HMSO, London.

*108 Nature Conservancy Council (1977). *Nature Conservation and Agriculture*. Nature Conservancy Council, London.

109 Nature Conservancy Council (1980). Amberley Wild Brooks, In *Nature Conservancy Council Fifth Report*, pp. 46–7. HMSO, London.

110 Nature Conservancy Council (1980). Lowland agricultural habitats (Scotland): airphoto analysis of change. By I. Langdale-Brown, S. Jennings, C. L. Crawford, G. M. Jolly and J. Muscott. NCC, CST Report No. 332.

111 Nature Conservancy Council (1981). *Loss and Damage to SSSI in 1980*. Unpublished report by Professor N. W. Moore, Chief Advisory Officer. Nature Conservancy, London.

112 Nature Conservancy Council (1982). Damage to SSSIs: pressures on sites of importance for nature conservation. In *Nature Conservancy Council Seventh Report*, pp. 18–20, 22–28. Nature Conservancy Council, London.

113 Nature Conservancy Council (1983). *SSSIs: What you should know about Sites of Special Scientific Interest*. Nature Conservancy Council, London.

*114 Nature Conservancy Council (1984). *Nature Conservation in Great Britain*. Nature Conservancy Council, London.

115 Nature Conservation in Great Britain (1943). *Report by the Nature Reserves*

Investigation Committee, Memorandum No. 3 (1943). Society for the Promotion of Nature Reserves, London.

116 Newton, I. (1984). Uses and effects on bird populations of organochlorine pesticides. In *Agriculture and the Environment*, ed. D. Jenkins, pp. 80–8. Natural Environment Research Council, Cambridge.

*117 Nicholson, M. (1967). *The System. The Misgovernment of Modern Britain*. Hodder & Stoughton, London.

118 Olsson, M., Johnells, A. G. and Vaz, R. (1977). DDT and PCB levels in seals from Swedish waters. The occurrence of aborted seal pups. Proceedings from the Symposium on the seal in the Baltic, Lidingö, Sweden, June 4–6, 1974. *Report from the National Swedish Environment Protection Board*. PM 900.

119 Parslow, J. L. F. and Jefferies, D. J. (1973). Relationship between organochlorine residues in livers and whole bodies of guillemots. *Environ. Pollut.*, 5: 87–101.

120 Perring, F. H. and Farrell, L. (1977). *British Red Data Books: 1, Vascular Plants*. Society for the Promotion of Nature Conservation, Lincoln.

121 Perring, F. H., Sell, P. D. and Walters, S. M. (1964). *A Flora of Cambridgeshire*. Cambridge University Press.

122 Pests Act (1954). HMSO, London.

123 Peterken, G. F. and Harding, P. T. (1975). Woodland conservation in eastern England: comparing the effects of changes in three study areas since 1946. *Biol. Conserv.*, 8: 279–98.

124 Pollard, E. (1968). Hedges IV. A comparison between the Carabidae of a hedge and field site and those of a woodland glade. *J. Appl. Ecol.*, 5: 649–57.

*125 Pollard, E., Hooper, M. D. and Moore, N. W. (1974). *Hedges*. Collins, London.

*126 Polunin, N. (ed.) (1973–). *Environmental Conservation*. The international journal devoted to maintaining global viability through exposing and countering environmental deterioration resulting from human population-pressure and unwise technology. Elsevier Sequoia, Lausanne.

127 Potts, G. R. (1980). The effects of modern agriculture, nest predation and game management on the population ecology of partridges (*Perdix perdix* and *Alectoris rufa*). *Adv. Ecol. Res.*, 11: 1–79.

128 Prestt, I. (1965). An enquiry into the recent breeding status of some of the smaller birds of prey and crows in Britain. *Bird Study* 12: 196–221.

129 Prestt, I., Cooke, A. S. and Corbett, K. F. (1974). British amphibians and reptiles. In *The Changing Flora and Fauna of Britain*, ed. D. L. Hawksworth, pp. 229–54. Academic Press, London.

130 Prestt, I. and Jefferies, D. J. (1969). Winter numbers, breeding success and organochlorine residues in the Great Crested Grebe in Britain. *Bird Study* 16: 168–85.

131 Prestt, I., Jefferies, D. J. and Moore, N. W. (1970). Polychlorinated biphenyls in wild birds in Britain and their avian toxicity. *Environ. Pollut.*, **1**: 3–26.

132 Prestt, I. and Mills, D. H. (1966). A census of the Great Crested Grebe in Britain 1965. *Bird Study*, **13**: 163–203.

133 Prestt, I. and Ratcliffe, D. A. (1972). Effects of organochlorine insecticides on European bird life. *Proc. Int. Orn. Congr.*, **15**: 486–513.

*134 Rackham, O. (1976). *Trees and Woodland in the British Landscape*. Dent & Sons, London.

135 Ratcliffe, D. A. (1963). The status of the Peregrine in Great Britain 1963–4. *Bird Study*, **10**: 56–90.

*136 Ratcliffe, D. A. (ed.) (1977). *A Nature Conservation Review*. Cambridge University Press.

*137 Ratcliffe, D. (1980). *The Peregrine Falcon*. T. & A. D. Poyser, Calton.

138 Ray, J. (1660). *Catalogus Plantarum circa Cantabrigiam nascentium*. John Field, Cambridge (for translation see Ewen and Prime above).

139 Relton, J. (1972). Disappearance of farm ponds. *Monks Wood Experimental Station Report for 1969–1971*, 32. The Nature Conservancy.

140 Reynolds, C. M. (1979). The heronries census: 1972–1977, population changes and a review. *Bird Study*, **26**: 7–12.

141 Robins, M. and Bibby, C. J. (1984). Dartford Warblers in 1984 Britain. *Brit. Birds*, **78**: 269–80.

*142 Rudd, R. L. (1965). *Pesticides and the Living Landscape*. University of Wisconsin Press, Madison, USA.

143 Several Hands (1726). *A Natural History of Ireland in Three Parts*. George Grierson, Dublin.

144 Sharrock, J. T. R. (1976). *The Atlas of Breeding Birds in Britain and Ireland*. British Trust for Ornithology, Tring.

*145 Sheail, J. (1971). *Rabbits and their History*. David & Charles: Newton Abbot.

*146 Sheail, J. (1976). *Nature in Trust. The History of Nature Conservation in Britain*. Blackie, Glasgow.

*147 Sheail, J. (1985). *Pesticides and Nature Conservation. The British Experience 1950–1975*. Clarendon Press, Oxford.

*148 Shoard, M. (1980). *The Theft of the Countryside*. Temple Smith, London.

149 Sladen, W. J. L., Menzie, C. M. and Reichel, W. L. (1966). DDT residues in Adélie Penguins and a Crabeater Seal from Antarctica. *Nature*, Lond., **210**: 670–3.

150 Steele, R. C. and Welch, R. C. (1972). *Monks Wood. A Nature Reserve Record*. The Nature Conservancy, Natural Environment Research Council, Huntingdon.

151 Taylor, J. C. and Blackmore, D. K. (1961). A short note on the heavy mortality in foxes during the winter 1959–60. *Vet. Rec.*, **73**: 232–3.

152 Taylor, S. M. (1983). The common bird census. In *Enjoying Ornithology*, ed. R. Hickling, pp. 59–67. T. & A. D. Poyser, Calton.

153 Thomas, A. S. (1956). Biological effects of the spread of myxomatosis among rabbits. *Terre Vie*, **103**: 239–42.

154 Thomas, A. S. (1963). Further changes in vegetation since the advent of myxomatosis. *J. Ecol.*, **51**: 151–86.

155 Thomas, J. A. (1980). The extinction of the large blue and the conservation of the black hairstreak butterflies (a contrast of failure and success). *Annu. Rep. Inst. Terr. Ecol.*, **1979**: 19–23.

*156 Thompson, H. V. and Worden, A. N. (1956). *The Rabbit*. Collins, London.

157 Tubbs, C. R. (1967). Numbers of Dartford Warblers in England during 1962–66. *Brit. Birds*, **60**: 87–9.

158 Tubbs, C. R. (1974) *The Buzzard*. David & Charles, Newton Abbot.

159 Turtle, E. E., Taylor, A., Wright, E. N., Thearle, R. J. P., Egan, H., Evans, W. H. and Soutar, N. M. (1963). The effects on birds of certain chlorinated insecticides used as seed dressings. *J. Sci. Fd Agric.*, **14**: 567–77.

160 Walters Davies, P. and Davis, P. E. (1973). The ecology and conservation of the Red Kite in Wales. *Brit. Birds*, **66**: 183–224, 241–70.

*161 Ward, B. and Dubos, R. (1972). *Only One Earth. The Care and Maintenance of a Small Planet*. André Deutsch, London.

*162 Warren, A. and Goldsmith, F. B. (eds) (1983). *Conservation in Perspective*. John Wiley & Sons, Chichester.

163 Way, M. J. and Banks, C. J. (1964). Natural mortality of eggs of the black bean aphid *Aphis fabae* Scop., on the spindle tree *Euonymus europaeus* L. *Ann. appl. Biol.*, **54**: 255–67.

164 Webb, N. R. and Haskins, L. E. (1980). An ecological survey of heathlands in the Poole basin, Dorset, England in 1978. *Biol. Conserv.*, **17**: 281–96.

165 WHO (1980). *Sixth Report on the World Health Situation*. World Health Organisation, Geneva.

166 Wigglesworth, V. B. (1945). DDT and the balance of nature. *Atlant. Mon.*, **176**: 107–13.

167 Wildlife and Countryside Act (1981). HMSO, London.

168 Wilson, R. (1984). The Mountain Gorilla project: progress report No. 6. *Oryx*, **18**: 223–9.

169 Working Party on Precautionary Measures Against Toxic Chemicals Used in Agriculture (1951). *Toxic chemicals in agriculture*. HMSO, London.

170 Working Party on Precautionary Measures Against Toxic Chemicals Used in Agriculture (1953). *Toxic chemicals in agriculture: residues in food*. HMSO, London.

171 Working Party on Precautionary Measures Against Toxic Chemicals Used in Agriculture (1955). *Toxic chemicals in agriculture: risks to wild life*. HMSO, London.

172 Wyllie, I. (1976). The bird community of an English parish. *Bird Study*, **23**: 39–50.

173 Yemm, E. W. and Willis, A. J. (1962). The effects of maleic hydrazide and 2,4, dichlorophenoxyacetic acid on roadside vegetation. *Weed Res.*, **2**: 24–40.

INDEX

*

(Figures in bold type refer to illustrations.)

research (*continued*)
  on pesticides, 148–81
  on PCBs, 195–9
  value of link between research and
    conservation, 216–17, 221
resistance to disease, 124
  to pesticides, 6, 146, 205, 207
Review of the Persistent Organochlorine
    Pesticides (1964), 192
rice, 207
Robertsbridge, Sussex, 138
Rockrose, White (*Helianthemum
    apenninum*), 77, 93, 124
rodenticides, 142, 173
  warfarin, 6, 173
Rodney Stoke NNR, Mendips, Somerset,
    71, 75, **75**, 80
Romans, the, 11–12, 15, 78
Romney Marsh, Kent, 13, 31
Rothschild, Hon. Miriam, 137, 140
Rothschild Report, 216
Royal Institute of Chartered Surveyors,
    107
Royal Society for Nature Conservation,
    107
Royal Society for the Prevention of
    Cruelty to Animals (RSPCA), 138
Royal Society for the Protection of Birds
    (RSPB), 29, 37, 107, 168, 172, 241,
    259
Rudd, Professor Robert, 154, 275
Ruff (*Philomachus pugnax*), 16, 238
Rwanda, 222
Rye-grass, Italian (*Lolium corniculatus*), 8,
    21, 23, 28, 178, 236

SSSI (Sites of Special Scientific Interest),
    bufferland for, 100–1
  climate change and, 96
  consultation with NCC over, 59–60,
    64–6
  damage to, 60–6
  development of the system, 29, 58–60,
    114, 116–17
  economic significance of, 90–1, 100
  function of, 59, 67, 85–6, 91–2, 102–3,
    236, 238, 240
  grading of, 90
  importance of, 59, 66, 85, 101
  management agreements for, 265
  notification of, 59, 86–8, 246
  ownership of, 113
  rivers and, 168
  selection of, 85–102, 113, 216
St Abbs Head, The Borders, 169
St Aldwyn, Lord, 140
St Bees Head, Cumbria, 169

St Ives, Huntingdon, Cambridgeshire,
    161
Sanarelli, G., 122
Sandwith, Noel, 76, 78
Saxons, the, 12–13, 19–20, 22
Scandinavia, 50, 237
Schradan, 149, 153
Scientific Areas, 83, 86
Scolt Head NNR, Norfolk, 165, 170
Scotland, ix, 12–13, 30, 42, 44, 62, 70, 96,
    108, 169, 172, 174
Scots Pine (*Pinus sylvestris*), 36
Scott, Sir Peter, 222
Scottish Colleges of Agriculture, 116–17
Scrub control *see* herbicides
sea, 7, 73, 170, 183, 195, 214
Seal, 197, 250
  Grey (*Halichoerus grypus*), 225
seed-dressing, 18, 212, 230
serpentine rocks, 58
Shag (*Phalacrocorax aristotelis*), 169
Shearwater, Manx (*Procellaria pufffinus*),
    203
sheep, 20, 132, 174, 208
  grazing by, 89
Shelduck (*Tadorna tadorna*), 74
Shell, 151
Shoard, Marion, *The Theft of the
    Countryside*, 243, 275
Shooting Times, 137
Shoveler (*Spatula clypeata*), 30, 238
Shrike, Red Backed (*Lanius collurio*), 37
Silent Spring see Carson, Rachel
Silsoe, Conference (1969), 105–7, **106**
simazine, 157
Skokholm, Dyfed, 123
Skomer (*now* NNR), Dyfed, 125
Skylark (*Alauda arvensis*), 14
smelting, zinc, 200, 203
Snail, Bulin (*Ena montana*), 77, 77, 79, 210
  Rough Mouthed (*Pomatias elegans*), 77,
    77
Snake, Smooth (*Coronella austriaca*), 34
Snipe (*Capella gallinago*), 30, 238
soil animals, 155
Solomon's Seal (*Polygonatum multiflorum*),
    79
Somerset River Board, 74
Soviet Union, 164, 197, 206, 237, 245
space, exploration of, 226
Sparrow Hawk (*Accipiter nisus*), 171–2,
    174–7, 205, 211, 214
Sparrow, House (*Passer domesticus*), 109
species, common, 92, 103, 109, 152
  number of, 6, 226
  rare, 14, 16, 29, 93–4, 98, 103, 115, 152,
    161, 205, 238